Riding

the

Tiger

Open
Space
Technology

**Doing Business
in a
Transforming
World**

The
InterActive
Learning
Organization

Harrison Owen

Riding the Tiger

Doing Business
in a
Transforming World

by

Harrison Owen

ABBOTT PUBLISHING
POTOMAC, MARYLAND

First published 1991
Second Printing Fall 1993

ABBOTT PUBLISHING
7808 River Falls Drive
Potomac, Maryland 20854
U.S.A.
301-469-9269

Printed in the United States of America

Library of Congress Catalog Card Number 91-076873
ISBN 0-9618205-2-7

Dedicated to my children Cameron, Amy, Barry, Christy and Harrison. And to their children, the first one of which is Becky. And to all children everywhere, for they shall inherit the Earth. For better or worse.

Also by Harrison Owen

Spirit: Transformation and Development in Organizations

Leadership Is

Open Space Technology: A User's guide

Table of Contents

Chapter I Riding the Tiger to Somewhere 1

Chapter II Chaos 11

Chapter III Chaos and Learning 30

Chapter IV The Process of Transformation 49

Chapter V The Journey Continues:
The Four Stages of Organizational Life 71

Chapter VI The Stages of Transformation: Where are we? 88

Chapter VII The InterActive Learning Organization . . . 116

Chapter VIII Let Go, Let It Happen,
Make It Better 148

Appendix A BRIEF USER'S GUIDE TO
OPEN SPACE TECHNOLOGY 182

PROLOGUE

Were this still the 18th century, I might well begin this prologue with the words *Dear and Gentle Reader*. With the vision of hindsight, those bygone days appear almost bucolic, a pastoral time far from the mad rush of the late 20th, soon to be 21st Century. But hindsight is not always clear sight, and what may appear to us now as a gentler time was in fact an era of stupendous change and transformation. The industrial age was ready to burst upon us, and elements of the human spirit soared on the euphoria of reason and science. The world was ours to rationally control and scientifically manipulate for profit and glory. Railroads soon rushed west in the United States, and industry mushroomed around the globe. There was a sense of manifest destiny unrolling before humankind.

Since then, we have had a number of surprises on the way to the future. And not all of them have been pleasant. New ages have been announced, only to be overcome and surpassed by succeeding New Ages. But through it all, there has been an abiding confidence in the human capacity to manage and control this small piece of solar driftwood we all call home, and if not the planet, then at least our own business. That was our job, our God-given task, the meaning of "being human." Or so we thought.

Some time in the past 10 years that abiding confidence appeared not so abiding, and even less confident. We are facing the fruits of our labors, and some (although certainly not all) seem less than fruitful. The overt sign is the imminent ecological disaster. We have fouled our own nest, and are about to face the consequences. The Industrial Age, along with the consciousness of humankind that gave birth to it, seems to be running out of steam.

There are those who feel that we need only to do better what we have been attempting to do all along. If our approach has been to control, our problem is that we have controlled less than well. If chaos is breaking out in our midst, we must increase our capacity to control and create order.

Others, and I count myself among them, feel that it may be time to take another look at who and what we are. If being in control is synonymous with being fully human, we are indeed in desperate straits. We need go no further than the front page of the morning paper to document the fact that nobody is in charge. Not the heads of state, nor the heads of industry. Nobody.

But perhaps we never were in control. And perhaps the notion of control as we have developed it was only a figment of our imagination. At best, a passing stage useful for our development. At worst, vaulting egotism, hubris, and pride leading to our destruction. Doing better what we have done before will only get us what we presently have. Only more so. But what next?

This book is more a travelogue, a story, than a prescription for the necessary steps to a better humankind. I believe we are on a journey, and the far mountains are coming into view. Already we have intimations of what the new landscape might look like which may provide the necessary vision to lead us on the way. But the issue of the moment is the conduct of the journey, the pursuit of the quest. We will reach our destination soon enough. Or we won't. The deciding factor is how we make the trip.

Dear and gentle reader, I invite you to listen to my tale. Where is it useful, make it your own. Where it displeases, make it better. It will, perhaps, provide some useful reading for the journey.

Harrison Owen
Potomac, Maryland
Fall 1991

Chapter I

Riding the Tiger
to
Somewhere

    Riding the Tiger became an image for the age and the title of this book while I stood before a small shop in Bombay. Displayed prominently in the window was a figurine of what I took to be Lord Shiva mounted on a tiger. As it turned out, it was Shiva's consort, but there was no mistaking the tiger.

    Even in the West, we know enough about tigers to be aware that riding them must be done carefully. Jumping off is a sure invitation to lunch. Trying to stop the beast is not advisable, for we know all about getting a tiger by the tail. And last but not least, the thought that we might somehow control our mount is ludicrous.

    For the Indians, who have much more intimate experience with the creature, the tiger has always been the symbol of deep power. Rather like the dragon further to the east, the tiger symbolizes the fundamental forces of existence, which must be acknowledged, honored, and respected if life is to be possible and meaningful.

It takes no crystal ball, or advanced degree, to know that we are all riding the tiger to somewhere. It is not simply a question of "more and faster" as Alvin Toffler described in *Future Shock*.[1] In addition, there is a fundamental change in *kind and quality*. The times definitely are changing, no longer in small incremental jumps, but in quantum leaps. Many words have been used to describe our present condition, but one seems appropriate: transformation. Somehow the very nature of our existence, or at least the way we perceive that existence, is radically different, and becoming more so.

Not too long ago, the word *transformation* saw only limited use in religion and psychoanalysis. The root meaning: a fundamental change of state, the passage from one way of being to another. For individuals this passage might be negotiated with the assistance of an analyst or guru (teacher). Now it would appear that we are all on this journey, although the identity of our teacher remains somewhat obscure.

Scarcely a day passes without the popular press acknowledging the transformational nature of our times. The *Wall Street Journal* announces that some corporation, faced with imminent disaster, is in the process of transforming the way it does business. Even the stately *New York Times* acknowledges the transformation of our world amid "All the news that is fit to print." Within the past few years, we have seen the Berlin Wall fall, Saddam Hussein attack and retreat, the advent of the "New

---

[1] Toffler, Alvin, <u>Future Shock</u>, Bantam Books, 1970.

World Order" (thank you, George Bush), the dissolution of the Soviet empire, Gorbachev's ouster and return, and the end of apartheid in South Africa. And that was just the past two years and only on the geo-political level. Add the race of technology, a growing realization of pervasive eco-disaster, and even the word "transformation" seems rather palid for the task at hand.

Use whatever word you like, but things are surely different. Under the circumstances, "riding the tiger" seems an appropriate image. We are all riding the tiger to somewhere, and the central questions are: Where are we going and how shall we make the journey?

It is the purpose of this book to deal with these questions. I make no pretense that mine is the only answer, or even the best answer. Actually, I propose to tell a story, blending elements of my own experience and the experience of many others. And of course if you do not like my story, you are invited to tell your own.

Storytelling may seem a weak reed upon which to place much confidence in times such as these. We have all been instructed from our youth that stories, by definition, are frivolous, suitable only for childish amusement. As adults, we must deal with cold, hard, objective facts, and leave the storytelling to kindergarten teachers.

That is what we have been taught, but I think it is a great loss. The storytellers of humankind are the myth makers, who weave together the strands of collective experience to create a cognitive fabric that makes sense. They create the context within which the details of living have meaning, or in Kuhn's terms,[2] the paradigm, which gives definition to the cold objective facts. No story, no facts. Different story, different facts.

If storytelling is still not sufficiently serious for your consideration, perhaps you would prefer *theory* — which the *American College Dictionary* defines as a "more or less verified or established explanation accounting for known facts or phenomenon." Translated to English, we might say a theory is a likely story — and once again we come back to storytelling.

I am, unashamedly, a storyteller. And what you are about to read is a story. Whether or not it is a good story remains to be seen, and for you to judge. If effective, the utility of the story will appear in its capacity to create a context of understanding in which our tiger ride begins to make sense, both in terms of where we are going, and how we might survive the trip.

One additional confession. I do not have The Plan, and certainly not one spelled out to a level of detail necessary for implementation. But a good story is never complete. Were it so, there would be no room for the imagination. You will find plenty of room for your imagination, and I invite you to use it liberally. Who knows, we may co-creatively produce a really good story.

The heart of my tale is that our wild tiger ride is nothing more, or less, than the current phase of the evolution of human

---

[2] Kuhn, Thomas, <u>The Structures of Scientific Revolution</u>, The Univ. of Chicago Press, 1962.

consciousness, the continuing saga of the maturation of the species. In a word, we are suffering growing pains. The good news is that we have been down a path like this before, not quite the same to be sure, but sufficiently similar to give us clues as to what might be expected. The even better news is that if we make the passage, we will more fully realize our essential human potential. We are growing up.

There is some bad news too. First of all, there are no guarantees. We may not make it. However, the fact that we have continued on this planet as long as we have, despite the vicissitudes of the journey, gives some small hope that we will reach a safe haven. It would be a mistake, however, to think of this safe haven as the end of the road, for I rather expect that we have more adventures in store for us.

The second piece of bad news is that the journey will not be without pain. There is no free lunch. Leaving our present state, in order to achieve our future state, will not be accomplished without genuine loss, and no small amount of anxiety. We can, however, make the journey infinitely more painful by refusing to take the trip, seeking to alter the course, or getting off in the middle of the ride. Those who ride the tiger need to understand that the tiger is in charge. Attempting to alter that situation can be very painful.

THE STORY IN BRIEF

To begin with a statement of the obvious, we are in chaos. This condition is particularly painful for those of us who have made a fetish out of being in control. The truth of the matter is that we never were, except in a rather limited way, but somewhere along the line control became the *sine qua non* for meaningful life, at least in the West. Western managers prided themselves on their capacity to control, and to be out of control was *prima facie* evidence of failure. And, of course, the opposite of control was (is) chaos. Well, we are now in chaos.

For much of world history, the appearance of chaos was not viewed automatically as a total disaster. While never pleasant, chaos also has a positive side. Drawing on some of our more ancient traditions, in addition to modern chaos theory, we will suggest that life without chaos is no life at all. Indeed, it is chaos that provides the growing spaces, the open spaces in which life evolves. The significant gift of chaos is to create the "differences that make a difference" (Gregory Bateson[3]), through which we learn.

Our learning concerns not just the facts and figures of everyday life, but rather the high learning indicative of real creativity: making something where nothing existed before. And also the deep learning, where we come face to face with what we really are, our essence so to speak.

---

[3] Bateson, Gregory, <u>Steps to An Ecology of Mind</u>, Ballantine Books, 1972.

Speaking of essence takes us to the core of the story: Spirit. This book is about Spirit, or more exactly the journey of Spirit. I take it as given that Spirit is the most important thing, for individuals and organizations. If this statement appears radical, even off the wall, I suggest it is merely the common knowledge and experience of every coach, champion, and sensitive executive. Everybody knows that you can have all the external manifestations of success, but unless the Spirit is present and willing, not very much happens. Capital, plant, facilities, product, game plan, elegant stadiums all amount to nothing without Spirit. Indeed, with the proper Spirit, amazing things can happen with none of the above.

The statement is also very much in line with all of the major world traditions. By different names, and in different ways, the ancient and continuing stories of humankind point to the centrality of Spirit as the deep core of reality. Of course, such historical testimony does not make it true, but that is the story.

We will take a close look at the journey of Spirit, considering what happens along the way. If it looks a lot like death and birth, that is not happenstantial. Perhaps it is no more than analogue and metaphor, but somehow the universal processes of ending and new beginnings have much to tell us as we ride the tiger towards whatever it is that we are becoming. It turns out that the process of Spirit's transformation, like the processes of birth and death, cannot be fundamentally altered. However, being aware of what comes next can ease the anxiety level, and in truth, there are a number of positive things to be done to smooth the way.

Next we examine where we have come from and where we might be headed. Borrowing liberally from ancient traditions and modern organizational theory, the way stations of our passage are described. Beginning with the *ReActive Organization*, in the primal world of the entrepreneur, the course is plotted to our present position as *ProActive Organization*, the rational, scientific, and well-planned operation which is at once powerful and insufficient for the emerging environment.

Once more we are preparing to make a jump, indeed the jump may already have been made. On the yonder side lies the *InterActive Organization*, unique in its potential for open, responsible communication with the environment at large. In prospect (before the leap), the InterActive Organization may appear counter-intuitive, irrational, and strange. But in retrospect, it will turn out that the InterActive Organization possesses its own rationality in which intuition is not counter to anything, but an essential of doing business.

At the present moment, as we either prepare for, or recover from, the leap into the unknown, it may seem superfluous to consider yet another leap. But I suspect that we have by no means reached the end of the road, and indeed the best is still ahead. My word for it is the *Inspired Organization*.

If the InterActive Organization is only now coming into view, the Inspired Organization remains cloaked in mystery, but already there are some signs of what it might be like, revealed in those momentary experiences of superior performance. Beyond technique, beyond clocks and constraints, there apparently lies a realm of human endeavor which is all Spirit. Perhaps it is, and will remain, the peculiar domain of poets and other romantics, but we might at least contemplate a time when *the beauty we love is what we do.* (Rumi)

Back to the present, or at least the very immediate future, for a hard look at the InterActive Organization, what it is, how it might work, and most important, how we get there. It turns out that the InterActive Organization is actually our new friend, the *Learning Organization*, although the curriculum and method of procedure go far deeper than some of the present discussions might indicate. Current conversations about Learning Organizations often leave the impression that the path forward lies in imposing order on our chaos through endless courses, seminars, and workshops, laid on our already overburdened organizational life. The end product looks suspiciously like a hybrid between the corporation and the university, with minimal changes in either. If that is the learning organization, I suspect that it will be of little utility in the emerging environment.

The Learning Organization is rather a different beast. In it chaos is embraced as the creator of the Open Space, in which High Learning may take place. And learning is not something we do now and then in some special time and place. Rather, every moment becomes a learning moment, as the distinction between *learning* and *doing* progressively disappears. Pollyannaish perhaps, but nevertheless a sure cure for a bad case of future shock. More important, the alternatives are not very attractive.

Getting from here to there may be accelerated by a practical, new technique for fostering the growth of InterActive Learning Organizations. We call it *Open Space Technology* (OST). Developed over the past several years, Open Space Technology has the demonstrated capacity to enable groups in excess of 400 people to self-organize large, task-oriented meetings in less than one hour, and then self-manage the process to achieve positive and substantive outcomes.

While the overt statistics may be impressive, the actual impact is even more so. In Open Space, the collective experience is not that of talking *about* Learning Organizations, but of having *become one.* Grounded in that experience, the group then has the opportunity to do it all better, and further, to spread the benefits into the everyday work environment. And no matter what, the group can never say it didn't happen.

This last point is important, for the jump between our present way of doing business in the ProActive Organization, to the new one as an InterActive Learning Organization, is not an orderly, linear progression. On a logical basis, you cannot get there from here. The only way is to take the leap, and then figure out what is going on upon arrival. OST may provide the jump-off point and a preview of coming attractions.

Chapter II

Chaos

If there is a single sacred word in the culture of most of our organizations, that word is *control*. When we have it, we are in good shape, and in its absence disaster is a short step away. As managers, we have been trained to control, and control is the prime attribute designating high-quality management. The centrality of control is not usually stated so blatantly, but it is never far from the surface. According to the old dictum, the good manager makes the plan, manages to the plan, and meets the plan. And the essence of all of that is control. Close, tight control.

We presently find ourselves in rather strange circumstances. It remains relatively easy to make a plan, for after all, we control the pen, paper, and computer. But insuring that the plan, once made, will have any relevance past the drying of its ink, is no easy task. Sure as the sun rises, some unpredicted event will shatter our best efforts. Saddam marches south throwing the world markets into confusion. Were we in the airline business, we would suddenly find ourselves flying empty 747s, which is unprofitable, and definitely not according to plan. Then again our business might be defense, and all our plans based on the implacable, undying animosity between the West and the East. Now there was something you could count on — until the fall of the Berlin Wall. Hard days for plan makers, and all those other folks who place high value on being in control.

But what are the options? Somewhere along the line we came to the conclusion that the only alternative to control was being out of control. And we all know what means. Chaos!

In the good old days (whenever they were), events moved at a stately pace, allowing us to make our plans with some reasonable hope of completion. And indeed, we often looked forward to a little chaos just for added spice. After all, when one plan was destroyed, there was always the opportunity to solve the problem and create a new one. We were a generation of "problem solvers," who prided ourselves on our ability to put things in order and apply the proper "fixes." It now appears that the problem solvers of the world have more opportunity than they bargained for. Chaos is no longer a little spice added to the organizational stew. It has become our daily bread and butter. As Mikhail Gorbachev said, "We are already in a state of chaos." (*Washington Post*, fall 1990)

RESPONSES TO CHAOS

When chaos strikes, our responses are fairly predictable. The immediate reaction is to run. Presumably, there is some place beyond the chaos where we will be safe. Unfortunately, we are discovering again  what the navigators learned in the 15th century. Our world is round, and running will eventually bring us back to our starting point.

An alternative reaction might be termed the ostrich routine. As we all know, the ostrich, faced with inescapable danger, places the head in the sand, apparently believing that out of sight is not only out of mind, but out of danger. Whether or not ostriches actually do that I cannot confirm on the basis of personal experience, but I do know that many representatives of *Homo sapiens* exhibit the behavior. Occasionally, chaos, being chaotic, randomly skips on by, leaving the victim unharmed. This phenomenon is often taken as validation of the strategy, but I think most would agree, it is only dumb, blind luck.

A slightly more reasoned approach is to concentrate on short-term results. When all the world is chaotic, and chaos may strike at any point, it only makes sense to get in, get out, and get what you can — as quickly as possible. This strategy is neatly summarized in the old adage, "Live fast, die young, and have a good-looking corpse." While immediately appealing, and effective for the moment, the strategy leaves something to be desired when it comes to genuine institutional development, and the creation of wealth as opposed to making money. A classic example, raised to the level of high art (perhaps black art), is the sorry spectacle of the United States corporate scene in the 1980s. With corporate raiders and merger mania, money was made by the truckload, and little, if any, wealth created. No doubt we will be paying for that madness for a long time.

Last, there is the fervent hope, perhaps desperate plea, that someday we will get back to normal. Somehow, if we can hold out long enough, hide deeply enough — this too shall pass. At long last we will get back to those halcyon days when order ruled, plans could be completed, and control was possible. Perhaps.

SOMEDAY WILL NEVER COME

The hope for a return to normalcy is precluded by myriad factors. I mention only two. First, the state of the planet. Second, the *Electronic Connection*.

It is not my intention to deliver an impassioned plea for ecological reform, although that is certainly in order. Rather I merely wish to point to the present sorry state of the planet as a prime factor precluding any possible return to normalcy. Take whatever list of ecological disasters you wish, (present, imminent, or potential), and it is patently obvious to even the casual observer, that the base system, upon which all other systems stand, is badly out of whack, and showing every sign of becoming more so. Acid rain, global warming, depletion of the ozone layer, destruction of the planetary lungs (rain forests), toxic wastes, and so many more that is seems almost pointless to count them. Each contributes, and all conspire, to create the conditions under which we will never return to normal, and business as usual. For it was "business as usual" that got us into this mess.

None of this is news, but until fairly recently (and maybe even presently) the argument was made that ecological concerns were fine just as long as they didn't hurt the economy. The argument is usually put more bluntly — "You can clean up anything you like just as long as my job is not affected."

Understandable as this argument might be, especially from the point of view of those with jobs on the line, it is palpably shortsighted, and betrays a basic misunderstanding of the relationship between economy and ecology. A short lesson in Greek may help.

The word "economy" is composed of two Greek words, *oikos* which means house, and *nomos*, which means laws or rules. Together we get something like house laws, or house rules. In a word, the economy is simply the rules we have created to run an ordered household. These rules are arbitrary to a degree, even though some would understand them to have been written on the first day of creation. The situation is not unlike the house rules in a casino. Given a different casino or different circumstances, and you will have different house rules.

The word ecology starts out with the same Greek word, *oikos*. But the second part is different, significantly different. The last part of ecology comes from the Greek *logos*. *Logos* means word, but a very special sort of word. Those of you who have read the first chapter of the Gospel according to St. John may remember the opening line. "In the beginning was the word (*logos*), and the word was with God, and the word was God."

Leaving all questions of theology aside, it should be apparent that whatever else *logos* might mean, it is not just idle chatter, or silly words. It is rather closer to the "fundamental structure of reality" or even the "divine essence of all that is."

Thus, economy means "house rules," and ecology means, "the deep structure of the house," or maybe better, the foundation. As any householder will understand, if your foundation is shifting, or in bad shape, because the games you are playing in the living room are too rough, you simply have to change the game. Things will only settle down when a new game, with new rules, is established, congruent with the foundation.

Of course, some may argue that we have only to get the foundation back the way it used to be, and we can go back to playing the old game by the old rules. That is a wonderful idea, but I rather suspect that the shift has gone too far, and indeed it was the old rules that got us into the muddle in the first place. There are now too many of us (six billion, give or take a few), and the old rules were designed for a simpler, less crowded age. Adding insult to injury, we are all connected and know it. This brings us to the Electronic Connection.

THE ELECTRONIC CONNECTION

Not terribly long ago, the notion that our planet was a small electronic cottage appeared far out and avant-garde. However, two events in the recent past moved the notion from science fiction to everyday experience.

When the stock market crashed in 1987, those concerned with finance, which seemed to be just about everybody, had the strange experience of recognizing that the stock market was no longer just a physical place occupied by an elite group of people gathered on Wall Street (or any other location), rather it was the Great Computer Conference in the sky, which rolled on 24 hours a day. Every 16 hours or so, the Electronic Connection would touch down (rather like a tornado) in London, New York, Tokyo, or elsewhere, and then move on. The reality of global connection has become more than a philosophical thought.

Were the lesson of '87 lost on the world, the events of the Gulf War surely made the case. No matter where you went (and I happened to have been traveling from the United States, through Europe, to India at the time), everybody was glued to the tube, and in most cases, fixated on the same station, CNN. It was almost as if we were all one family, huddled in the living room, watching our favorite show. Although in this case, the show was real and generally unpleasant.

We are all connected and virtually instantaneously. When something happens in a far corner of the planet, we know it, and react. What all of this has to do with the impossibility of returning to "normal" is quite simple. As each part of the global family adjusts its rules (economy) to align with the new global deep structure (ecology), that adjustment forces other adjustments, and around we go again. More often than not, "my adjustment" is your "disaster" and vice versa.

Without the electronic connection, a rule change in one part of the planet might go undiscovered for years, and we would have a breathing space. But when that change hits the seven o'clock news, or appears in your electronic "in basket," the time for catching your breath becomes noticeably shorter. And so it goes — chaotically.

GUESS WHAT? THIS IS NORMAL. CHAOS IS A NATURAL PART OF LIFE

Slowly it is dawning on most of us: there is no going back, and what we now experience is normal. If this is so, then perhaps chaos is not antithetical to life, but rather a normal, natural, and possibly necessary, aspect of what it means to be alive.

This later thought may verge on the heretical, for have we not all been taught that the lack of order is the end of productive existence? Science, at least as we learned it in school, had one basic message. The universe is an orderly place, thus the scientific method and prediction are possible. From Newton onward, we have lived in a clockwork universe with a place and time for everything, and everything in its peculiar time and place. Were things to get out of order, it was the role of science to put it back together. Maybe.

Of course, there were some other aspects of the scientific endeavor that did not seem to play by the same rules. Subatomic physics, for example, found itself in world of randomness where indeterminacy was the rule, if a rule can be indeterminate. However, this may all have been an aberration. Did not the father of modern physics, Albert Einstein boldly proclaim that, "God does not play dice"? For Einstein, as for many of us, the thought of a fundamentally disorderly universe is appalling. Little storms and small disturbances to be sure — but chaos as a natural part of life?

19

A WORD FROM THE PAST

Actually, the thought that chaos is not only a natural aspect of life, but an essential and positive element, is not a new one. So far as I am aware, every major religious tradition has held this view. Of course, that does not make it true, but at least it may give us pause for thought.

For the Hindu, Shiva, the Lord of the Universe, is usually depicted with two faces. One of the faces is that of the creator. But the second is the face of destruction and chaos. The picture is relatively clear, the universe is the product of an alternation, or better, synergy of forces: order and disorder, chaos and cosmos.

From a different part of the globe comes a similar thought. The Taoist tradition of China places much weight on the yin and the yang. While often thought of as the male and the female polarities, there is in fact a deeper meaning. The yin and the yang can equally refer to the light and the dark, the forces of order and the break through of chaos. If life were all order there could be no evolution. Were it all chaos, there would be no continuance. It is only in the alternation between order and chaos that life progresses.

Perhaps the clearest expression of this whole complex of ideas in the Chinese context, is contained in the *I Ching*, the Book of Changes. Created some 3000 years ago, the book takes those who care to consult it through a journey of life's changes, from the moment of creative power to the very end. And in between, the subtle and not-so-subtle forces of creation and destruction play out.

The interplay of the powers of chaos and order, as an expression of the divine intent, finds its place also in Judaism. The sacred history of the people of Israel may be read as a guided passage through chaos and on to New Creation (to use the phrase from Jeremiah). From Egypt, into the chaos of the Desert, and on to the Promised Land. But note. The Desert is the antechamber to the Promised Land. The prophet Isaiah puts the thought quite directly when he says (speaking for God), "I create the light and make the Darkness. I create peace (*shalom*) and chaos (*tohu w' bohu*)."

And in Christianity too, those devotees who were raised with the children's prayer, "Gentle Jesus, meek and mild..." are sometimes shocked to hear the words from this same Jesus, "I came not to bring peace, but a sword..." Of course, those words are not so strange when it is remembered that Jesus met the end, which also was the fulfillment, of his earthly ministry in the chaos of the Cross. Naturally, Easter Sunday follows, but as many Christians are apt to forget, you cannot have Easter Sunday without Good Friday.

Is all of this true? Who knows, but that is the story, and it is a story that has been told in the community of humanity with remarkable consistency for a very long time. It is only in the recent past (since the dawn of the scientific age) that we have attempted to tell a different story in which disorder and chaos are banished from the universe as aberrant and fundamentally useless phenomena. Perhaps our new story is the aberrancy.

A WORD FROM THE PRESENT

In a curious sort of way, history seems to be repeating itself. For Science, or at least some part of Science, has now rediscovered chaos. Within the past dozen years, there has emerged, from a very broad spectrum of scientific disciplines, first a suspicion, and now something which looks remarkably like a coherent body of knowledge, all gathered under the umbrella of chaos theory. I will not describe chaos theory in detail, and refer the reader to an excellent book on the subject by James Gleich[1].

In a nutshell, the chaos theorists are proposing that not only is there a pattern in chaos, but chaos is useful. The pattern emerges upon consideration of the life cycle of any natural, open system. Actually, "natural open system" is redundant, for all natural systems are open.

An open system is in constant, unavoidable, interaction with its environment. Open systems are to be contrasted with closed systems, which turn out to be figments of our imagination, existing only as theoretical constructs. Even in the labora-

---

[1] GLEICH, JAMES, CHAOS: MAKING A NEW SCIENCE, PENGUIN BOOKS, 1987.

22

tory environment, where scientists do their level best to "close the system," and thereby control the (unwanted) variables, something always seems to get through. It may just be an aberrant neutron, with an impact so small as to be forgettable, but something always opens the can.

There is a lesson for managers in all of this scientific jargon, which we might note in passing. We have been treating our organizations as if they were closed systems which we might fully control, all under the heading of scientific management. The truth of the matter is that all systems are open, and most especially our organizations. Is it any wonder then that efforts to control inevitably meet with disappointment?

Now back to chaos. When you observe the process of a natural system, it is noted that the life cycle is punctuated by periods of order and chaos. Sometimes things go right, and sometimes we are in deep tapioca. There is no news here, but a definable, predictable pattern emerges. While one may not be able to say *when* this pattern will begin or end, *that* it will occur is assured.

The pattern divides into four stages. The first stage might be called Steady State with Development. Everything is going along fine, and getting better. The second stage is called Periodic Doubling — the meaning of which we will come to later. In the third stage, chaos appears, which means that all previous patterns are broken and predictability becomes a thing of the past. The final stage may have one of two forms: dissolution or renewal at a higher order of complexity. The meaning of dissolution should be obvious: everything falls apart, and it is over. Renewal at a higher order of complexity is the intriguing piece. Somehow this Open System gets itself back together, not as it was, but in a new (usually radically new) fashion, which is at once related to its past (it is still recognizably the same sort of thing), *and* in synergistic harmony with the environment.

For example, suppose that our Open System is a population of animals. Each year the mothers and fathers do what they are supposed to do, and the herd increases. We might say that it is stable and getting better, and predictably, given sufficient water and food, things will only improve.

But one year a very strange thing happens. For absolutely no observable reason, the number of births doubles. The next year, the number of births is halved. And so it continues for a

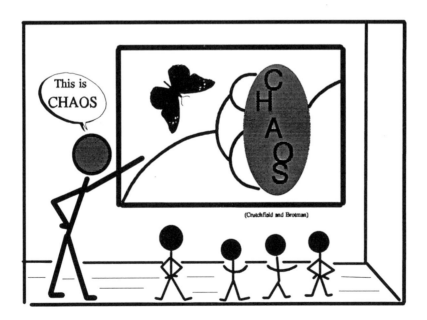

(Crutchfield and Brotman)

few years, doubling up and then doubling down (this is Periodic Doubling). After a time, and usually a very short time, any logic or rationale in the number of births totally disappears, and we have chaos. From that point onward one of two possibilities will come to pass. Either the herd will disappear from the face of the earth, or it will re-stabilize in some new functional pattern, more conducive to living in its environment.

The critical point is the onset of Periodic Doubling, and the critical question is, why did it occur? Here we must introduce the butterfly. One of the most profound discoveries of the chaos theorists is that *Open Systems have extreme sensitivity to early conditions.* Translated, that means that sometime in the early life of the herd something happened or didn't happen. At the time, this happening would have appeared so trivial as to be inconsequential. But somehow the impact of this happening was carried along in the life of the herd in a dormant state. Suddenly, for no apparent reason, the happening happens, the balance is tripped, and Periodic Doubling commences. Now back to the butterfly. It is part of the folklore of chaos theorists that a butterfly, flapping its wings in Thailand, will affect the weather system of California. Who knows whether it is true, and it is doubtful that the butterfly will ever be caught in the act. But that is the story.

So there is a pattern, and knowledge of that pattern can have actual utility. Gleich gives the example of electronic transmission. It has been known for years by those whose business it is to carry electronic messages around the world and beyond, that every now and then the whole system breaks down, and the message dissolves into noisy chaos. The presumption was that somehow the equipment utilized was less than perfect, and so a great deal of time and effort were devoted to enhancing that equipment. Cleaner switches, better wire, and so on. Doubtless this effort had some substantial payoff, but the basic problem remained. Every now and again, no matter what you did, things got chaotic. Then after a while, things cleared up, and it was back to business.

One fine day, it was noticed that the course of events was not quite as random as it seemed, and indeed they followed a course which was precisely what chaos theory would predict. Steady State, Periodic Doubling, Chaos, New State. The solution in this case was not to fight chaos, but to use the theoretical pattern as a predictor of the advent of chaos, and then prepare for the arrival. Concretely, this meant setting computers to detect the early elements of Periodic Doubling, whereupon they would switch over to a redundant system until such time as chaos had passed, at which point they could switch back to the original system. That is known as going with the flow in a rather novel way.

But there is more to dealing with the mystery of chaos than simply acknowledging its existence and preparing for the inevitable arrival. One might reasonably ask, what good is chaos?

WHAT EARTHLY GOOD IS CHAOS?

If chaos has a place in the natural order of things, it seems pertinent to ask, does it do any good? Has it any use, and if so — what?

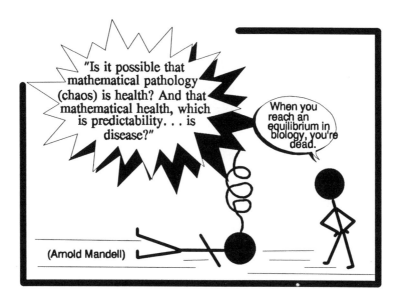

Arnold Mandell, quoted in Gleich's book, poses the question in an interesting, and provocative manner. "Is it not possible that mathematical pathology, i.e. chaos, is health? And that mathematical health, which is predictability... is disease?" He then makes the point pointedly. "When you reach an equilibrium in biology, you're dead."

The suggestion is that chaos represents the growth point in any system. Or in terms which we will be using rather extensively, chaos creates the Open Space in which the new can emerge. Obviously there are no guarantees here, for chaos can equally mark the end, in fact it always does. The central question is not about ending, but rather the possibility of new beginning. Chaos may therefore be the essential precondition for all that is truly new. No chaos, nothing new.

One of the unique aspects of chaos in my experience (and I suspect everybody else's) is *difference*. Whatever else may be true, the chaotic situation is different, unlike what preceded it and what follows. We may not like the difference, and indeed that difference may be downright painful. But there is no denying the difference.

Gregory Bateson teaches us that the perception of difference is the essence of learning. Or in his words, learning is "differences that make a difference." This deceptively simple phrase takes us in interesting directions, for it suggests, in the present context, that the function of chaos is to create the conditions under which real learning can take place.

While it may be true that chaos is part of the life story of all systems, our concern here is primarily human systems — practical concerns like businesses, governments, and other organizations, productive of the goods and services we as human beings require. Furthermore, the pressure of the moment (our Tiger Ride) makes it essential that we focus our attention not on the maintenance of what is, but the evolution of what must be if we as a species are to continue in some useful way on this planet. Learning, in its deepest sense, appears to be critical. Thus if chaos creates difference, and difference enables learning, may it not be that our nemesis is also, and simultaneously, our salvation?

So what good is chaos? Provisionally, let me propose that chaos creates the differences that make a difference, through which we learn.

> **Chaos creates the differences that make a difference, through which we learn.**

Chapter III

Chaos and Learning

The suggestion that chaos and learning are naturally linked, and more, that one forms the essential precondition of the other, may appear nothing short of lunacy. Do we not know, as only countless hours in the school room can teach, that learning requires order? What else does the teacher do but maintain order in the classroom so that learning may take place?

But do we not also know, as only a squirming fifth grader can know, that such order, even in mild doses (to say nothing of extreme application), can become exquisitely boring? Boring to the point that learning and boredom are often equated. Somehow, if we are not painfully bored, we can't be learning.

I can claim no expertise in the art and science of educating fifth graders, but I can bear testimony to my experience of that time under the iron hand of Mr. Birdsil. Mr. Birdsil's class was the very model of order. We sat in neat rows, spoke only when spoken to, and then only rarely. Mostly we listened while Mr. Birdsil pontificated on a variety of subjects, the impact of which was so minimal as to be insignificant. Occasionally, perhaps more than occasionally, the endless pontification would be interrupted by the abusive denunciation of some unfortunate who had fallen asleep. More usually, the denunciation was non-verbal, taking instead the form of a well placed shot with a blackboard eraser at the sleeping head.

I do, however, remember one significant event. I had a question, and following the required procedure, I raised my hand. When recognized, I began the ritual phrase, "Mr. Birdsil....." But instead of "Birdsil," what came out of my mouth was "Birdseed." I am sure the devil made me do it, for I have no consciousness at all of thinking such an outrageous thought. But there it was, hanging in the shocked silence of the awestruck classroom. Mr. Birdsil looked as if the devil himself had put in an unwanted appearance, and carefully laying his chalk and eraser on the desk, he strode with ominous purpose until he towered over me. His face was white with anger except for a little red spot on the tip of his nose, which apparently came from spirit of a different sort. Then he spoke — bellowed would be more accurate — "OWEN... what did you say?" And before I could even think of a reply, he struck me full force with an open hand in the face. I do remember that, indeed, that may be the only learning remaining with me from the fifth grade.

Say what you will, my encounter with Mr. Birdsil was different, and in that difference came learning. Not of the best sort perhaps, but learning nonetheless. Fortunately, the balance of my educational career was not a replication of the fifthgrade experience. I came to know that learning, excitement, enthusiasm, and inspiration could all go together. But mostly what I came to know is that learning takes place when difference is perceived. Gregory Bateson was right, the essence of learning is differences that make a difference.

If learning occurs when differences make a difference, surely we do not have to go all the way to chaos in order to achieve the desired effect. I think that is correct. At the same time, if we never go to chaos, or if we spend all of our time and effort avoiding chaos, then the possibility of genuine learning is limited indeed. However, some further definition and distinctions might be helpful.

First, chaos. In contemporary conversation we tend to reserve use of that word to those mega-buster situations where everything hits the fan. There is some value in this in that by using chaos only in situations of ultimate disaster, we can perceive our lives as being largely without chaos. And that is a great comfort. But there is also a loss.

The loss incurred by defining chaos in more or less extreme, absolute terms is to blind ourselves to a truth: everything is a question of scale,[1] and therefore a matter of perspective. Put rather more directly, my chaos can be your minor inconvenience, and vice versa. It all depends on where you sit.

For example, if you as company president conclude that one product line (out of one hundred) has become unprofitable, and therefore must be terminated, that is a minor, everyday

---

[1] The notion of "scale," and its graphic representation in fractal geometry, is very central to the work of the chaos theorists. For a fuller description of what is involved, see Gleich pp. 83.

business decision for you. However, if I am the maker of that product, having defined my past, present, and future in terms of its production, I will see the matter in a rather different light. For me it is chaos.

With such "absolute" definition we are forced to think in terms of order *or* chaos, when it is probably more appropriate to think of order *and* chaos, the two constantly in interaction at all levels of scale. In a word, there is never a moment when we do not have chaos heading toward order, or the other way around.

When we reserve the use of chaos for only those situations of ultimate disaster, we fail to see chaos as the everyday companion in life. Then, when the recognition of chaos becomes inescapable, it inevitably comes as a surprise, and usually a nasty surprise, for which we are not prepared. In the good old days, when events moved at less than their present meteoric rate, such an understanding was possible, and probably useful. But at the moment, when transformation laps transformation, it is necessary to become accustomed to chaos. And be prepared for it.

NORMAL LEARNING AND HIGH LEARNING

I am sure there is a place for the ordered classroom, no matter how much I may have found it unpleasant. That is the place for Normal Learning where we ingest all the details, facts, figures, and minutia needed to get along with life. All of that is necessary, but hardly sufficient. Unless there is some reasonable dose of what I would like to call High Learning, life moves along with monochromatic sameness.

The notions of Normal Learning and High Learning are borrowed from Thomas Kuhn (with some alteration). Kuhn actually talks about High Science and Normal Science. The former occurs at those moments of paradigm shift, when an old way of conceptualizing the world passes before a new one in what is usually a tumultuous, chaotic, event. Normal Science is what occurs after the new paradigm arrives. Cleaning up the territory so to speak.

It is but a small jump, I think, from High Science to High Learning, and probably only slightly different words for the same thing. High Learning occurs when chaos cracks the established order, permitting us to see some difference that makes a difference. We find ourselves on a quantum leap past, and through, what we knew before and on to a new way of perceiving the world. The chaos in question may be minimal as the world would see it, but it is sufficient to open vistas. The issue is always "sufficiency," and never some absolute quantity. After all, butterflies scarcely qualify as mega-events. Normal Learning is what we do after we make the trip. At some level it amounts to taking stock of the new territory.

Personally, I tend to be a High Learning addict, and others I know prefer the more ordered approach of Normal Learning. The obvious truth is that neither one alone will do the job, both are necessary. It is in those moments of High Learning that we experience the life-transforming events that take us individually, and as organizations, into breakthrough. New product, new thought, new reality, all are the gifts of chaos. But then, in Normal Learning, we tidy things up in order to take full advantage of our new perspective.

THE GIFT OF CHAOS — INNOVATION

Innovation is the gift of chaos, appropriated by High Learning, and made useful through Normal Learning. That rather bald statement encapsulates what I understand to be the central benefit of chaos for our organizations and businesses. Although extreme in appearance, that statement may also make some sense out of the strange phenomenon that all major breakthroughs (no matter how defined) always seem to occur by "mistake" — a polite way of talking about chaos. I know that is not the way things are *supposed* to happen, for we would all like to think that our advancement proceeds along an ordered course, well thought out in advance, and definitely according to plan.

The classic case was the discovery of penicillin and with it, the advent of the so-called miracle drugs. According to the story, we never would have had this wonder drug if Sir Alexander Fleming had washed his laboratory dishes. Fortunately, he made a mistake, and left a mess over the weekend. Upon his return he found a hairy, green substance growing in the dirty dishes. That was disturbing, but what caught his attention (a difference that made a difference) was that where the mold grew, bacteria did not. Naturally, prior training was necessary to be able to tell the difference between mold and bacteria, and also to perceive the lack of bacterial growth as significant. Normal Learning is important. However, it was the mess that catapulted Fleming from "more of the same old stuff" into genuine innovation.

Over the years, I have collected what can only be called anecdotal evidence from clients and colleagues concerning the circumstances surrounding real breakthroughs. The interesting thing is that absolutely none of them ever occurred according to plan. While I may have found only what I was looking for (which is usually the case), I am still searching for a break-through which happened the way it was supposed to.

The Birth of Fiberglas

Fiberglas, the discovery and major product of Owens/Corning Fiberglas (OCF), began with a mess. Shortly before World War II, OCF was seriously looking for other ways of using what it knew best, glass making technology. Up to that point it had largely been making bottles, but with the advent of plastic, it looked as if the bottle market might take a dive. So the search for new applications and products.

One fine day, their director of research decided that if a way could be found to weld glass blocks (the sort you build transparent walls with), that would be a new, marketable product.

I have never been clear exactly why he thought this was so, but he did. In any event, he summoned his research assistant, one Dale Kleist, and directed him to figure out the appropriate means.

Dale obediently assembled a pile of glass blocks, a gas torch, glass rods, and set about doing what he had been told. Unfortunately the fruits of his labor were not as envisioned. The harder he tried, the messier things got. As he melted the glass rods with the gas torch, preparatory to "welding," the force of the escaping gas blew the molten glass all over the floor — in long thin fibers. In a very short time, he had accumulated a considerable pile, and so far as he was concerned, the grand experiment was a disastrous mess.

As Kleist was reaching despair, the director returned to the scene of the crime. Kleist was prepared for the worst, but instead of loud denunciation for failure, the director was enraptured. What he saw in that mess was the tensile quality of the glass fibers, and FiberGlas was born.

The curious thing about this story is that forty years later, when I was consulting with a division of the company, virtually nobody remembered it, except for a few old timers. Even though that moment in OCF history required, as possibly never before, some useful examples of how to make an opportunity out of a mess.

The situation was a common one in the 1980s. The corporation had been attacked by a corporate raider, and management was doing its best to hold on. In the final round, management won, but it was a bittersweet victory. In order to meet the ransom the company sold businesses and closed facilities to the point that once robust annual sales of $4 billion shrank to a little more than $2 billion.

Even more critical was the fact that, even though not everybody lost their jobs (many folks went with the sold businesses), there was a very significant reduction in force. This meant that the business that remained had to be done with many fewer hands. It is a testimony to those who stayed that they put their best foot forward and rallied the company, but at tremendous cost. Fourteen-hour days, seven days a week, and at the end of six months, the folks were simply exhausted. There comes a point when you can't run any faster; you have to run smarter. But the options for smart running seemed limited indeed. It was a simple case of playing a new ball game by rules created in the halcyon days when money and staff were no problem.

And they had forgotten their story. Once upon a time, OCF had made opportunity out of a mess, virtue out of a mistake, new business out of a failed experiment. And doing all that again would be infinitely easier if they could remember having done it once before. No guarantees, of course.

How could they forget their story? The question really bothered me, and I have no certain answer, but I did notice a curious coincidence. Shortly before The Fall, OCF was proudly investing an incredible amount of money in the support of research. Millions of dollars went to maintaining a large research campus, home of 1200 people. Everything was carefully managed. Programs and systems piled on top of each other, all dedicated to insuring the relevancy of research to market needs. It was a well-oiled machine with no chaos allowed. There was, however, one small problem. According to local lore, the preceding 10 years of carefully managed research had produced absolutely no new products. Safer products, prettier products. But nothing new.

Given their recent history, it would have been very difficult to admit that everything had begun with a mess. And as a matter of fact, it is quite unlikely that, given the way they were doing research, Fiberglas would ever have been discovered. Rather, that mass of messy glass fibers would have been swept up, and Dale Kleist directed to take some new approach. After all, you have to stick with the plan. As for the story of Dale Kleist? Better forget the whole thing.

Breakthrough Technology

A research department of Dupont retained my services to assist them in achieving what they called "Breakthrough Technology." Apparently they saw the market taking some interesting, and not necessarily beneficial turns, and thought they should get ahead of the game. In the course of this assignment, I met with the directors of the several local laboratories, and asked them whether they had ever had any breakthroughs, on the grounds that if it had ever happened before, we would at least know what we were looking for.

After some thought, they identified six events that qualified. To this day, I am not entirely sure that they actually were, as each seemed to involve stranger ways of twisting molecules, none of which did I understand. But the directors were satisfied, and that was all that counted.

In order to get some sense of the importance of these breakthroughs, I asked what would be the profitability of their product line had these breakthroughs *not* occurred, and all agreed that the current bottom line results would not be pleasing.

My next question was a little rougher. How many of these breakthroughs, I asked, occurred according to plan, with the right people doing the right thing at the appropriate time, and place, all within budget? There was a very long pause. And the answer, when it came, seemed more than a little embarrassing. NONE.

Then I went to heart of the matter, and asked whether any of them had almost failed, not for technical reasons, but for other causes. There was an even longer pause, and eventually two candidates were named, but the reasons why remained unstated. I asked why, and a young manager answered almost sheepishly, "When we tried to manage them."

It struck me as both strange and sad that the only successes that these folks could identify occurred in spite of their best efforts to do what they were supposed to do. Further, failure loomed when they did their job.

Eventually the silence was broken by the same young manager who had last answered my question. He said, "Harrison, I think we are wasting a lot of our money and your time. All we have to do is do intentionally what it seems we are doing anyhow." I couldn't disagree with him, and that session marked the end of my assignment.

The simple truth of the matter was that these laboratory directors held a notion of research and innovation so predicated on orderly, programmed activity, that they simply couldn't recognize (without prodding) any significant event (read "breakthrough") which occurred outside of their expectations. Obviously they all "knew" that the breakthroughs had occurred, but their occurrence was treated as an aberrant phenomenon, an exception to the rule of ordered research. It turned out, of course, that the exception *was* the rule.

Forcing Mistakes to Happen

The story of the creation (invention) of Post-Its at 3M has been told so often as to have assumed a virtually unassailable position in current organizational mythology. The omni-present Post-It began with a terrible mistake, compounded by happenstance. The mistake was the creation of a glue that never quite set, and for an adhesive company, such a thing had better be forgotten. Then, happenstantially, a 3M employee, who sang in a local church choir, wished he had page marks that wouldn't fall out of his hymnal, but weren't permanently attached either. After all, a permanently gummed up hymnal would not be a thing of beauty, and you couldn't always sing the same hymns. As chance would have it, the two met, and the Post-It was born. Not without many trials and tribulations, to say nothing of total, initial corporate rejection, but it was born.

One of the young Turks who participated in this under-ground adventure, Tom Eckstein by name, was reflecting on their experience. Of the many things he learned and told us, one phrase sticks in my mind. "Learning," said Tom, "is forcing mistakes to happen, not just allowing them to happen." At the time, this suggestion seemed to take my notion of the necessity of chaos for creativity just a little bit too far. But when I pressed

Tom on the matter, he responded by reminding me how to build a good race car. You may start with the best available engine, but if you just leave it sitting in the garage, it won't get any better. Improve-

Learning is forcing mistakes to happen!

ment comes only when you take the beast out on the track, drive right through the "red line," blow the engine, and then build a better one. For Tom, breakthrough learning occurs truly when you force mistakes to happen.

Destroying the Bell System

Russell Ackoff tells a story, which I have used in an earlier book. But a good tale bears re-telling and it fits right here under the heading of forcing mistakes to happen in a grand manner.

It seems that the director of Bell Labs, that venerable research institution of AT&T which had produced so many breakthroughs, concluded that while present efforts were out-standing, and all according to plan, they were all, in effect, gilding the lily. A nasty rumor had surfaced in the industry that the future of telecommunications was digital, and all the present work was aimed at improving the existing analogue equipment (for example, the rotary dialing phone). The problem was you couldn't get there from here on a nice, straight road.

One morning, the director assembled his lab directors, giving no indication what the meeting was to be about. At the appointed hour, he entered the room, walked to the front without acknowledgment, and turned. In somber tones he said, "Gentlemen, last night the Bell System was destroyed. Your job is to build a new system within, or beyond, the range of technical possibility." And he left.

In one fell swoop, this leader of world-class research had forced a monumental mistake. He effectively cancelled the research agenda and reduced all the well-made plans to chaos.

But in the Open Space created, some remarkable things emerged. The TouchTone Phone, digital switching, and all the rest that we now take to be commonplace.

There is nothing approaching proof here, but in 10 years of asking, I have never found any person, presumably involved in innovative activities, who could remember any time that the breakthrough occurred according to plan. That may be faulty memory on their part, or faulty listening on mine. But that is the situation, and I believe it is significant.

Going to the Depths: The Ultimate Gift of Chaos

When chaos strikes, as it certainly shall, old things tend to pass away, and new things emerge, but in the Open Space between one and the other, there is a moment for seeing the really important things of life. While never comfortable, such a moment provides the essential opportunity for asking about meaning. In business, as in other aspects of living, we get so involved with the details that we tend to forget our purpose. Indeed, the details often become the purpose, and that is scarcely satisfying.

As the climax approached in the saga of Owens/Corning Fiberglas, senior executives found themselves in a maelstrom of activity. The only certainty was that nothing was certain, and they had to find some new way of doing business which would be acceptable to the banks, stockholders, customers, and employees. The name of the game was re-organize, not once, but quite literally dozens of times. Not that each organizational plan was published, but many were laid on the table as appropriate fit and function were sought. In the words of the chaos theorists, it was Periodic Doubling with a vengeance and total chaos was just around the corner.

In the midst of it all, there were periods of momentary respite, even silence, when it became possible, even mandatory, to ask the unaskable question: Why bother with it anyhow? What did it all mean? Asking such questions runs the risk of coming up with a troubling answer — there is no reason. And I suppose that can be beneficial.

On the other hand, the posing of the question can also create deeper opportunities. In the case of at least one OCF executive, I believe that occurred. After most of the dust had settled, this executive reflected on the situation as follows:

"We re-reorganized so many times that more than occasionally, I couldn't remember who we were. But the remarkable thing is that through it all we never lost our Spirit. However, if we had lost that, I think we would have lost it all."

The true Spirit of an organization often (usually) gets buried in the daily round of important things to be done. For a period of time, that situation is of no consequence, for after all, the business is being accomplished. But there comes a time when the state of the Spirit becomes of more than incidental concern. The initial signs are usually quite small and very forgettable. People just don't seem as involved and excited as they were in the "old days." At first, such observations are passed off as the nostalgic remembrances of the old-timers. But then it seems that something deeper may be involved. Organizational relationships become frayed, tempers snap. Arguments and backbiting break out for no apparent reason. The "zinger" replaces genuine humor in the corporate conversation. And the great "They" emerges as the source of all evil. They did this, They didn't do that, but nobody ever saw "They." Unless it was all those poor secretaries and junior accountants, who having been relegated to the backroom and cut off from the excitement of the business, now are left with the "administrivia" (trivial minutia), which is only trivial when it is not done.

Eventually more serious signs of a sagging Spirit surface. Vision goes, innovation slows, creativity is visible mostly by its absence. Customers go unlistened to, and quality is only something to talk about. A sagging Spirit is a weak Spirit which inevitably produces a sagging bottom line. For the truth of the matter is, Spirit is the bottom line.

Coming to this realization, or remembering it, is never a pleasant experience. For it usually occurs in the midst of chaos. At precisely the moment when we need every ounce of spirited participation that we can muster, the Spirit has apparently gone on vacation and off the job. That should come as no surprise, for nobody was taking care of the Spirit. Somehow, it just didn't seem to be an important thing to do.

In the good old days, back when things were normal, conversations about Spirit were best left to the more ethereal aspects of society (religion), while the rest of us got on with the important things of the moment, namely business. However, as the Tiger rides on, and transformation asserts itself as a continuing reality, concern for the Spirit of this place (our place of business) is more than incidental. The chaos of the moment has captured our attention and provided the Open Space in which we may (must) consider what is really important, and how to take care of the truly important things.

Chapter IV

The Process of Transformation

When chaos strikes, transformation begins. The Spirit of a place enters a critical phase with possible end results ranging from dissolution to genuine High Learning. While lady luck may have some minor role to play, the final result will depend largely on our capacity to understand the process and facilitate the outcome. Fully taking care of Spirit is an intentional and informed act.

THE NEW REALITY: WHAT GOES UP WILL COME DOWN

In the good old days, chaos and transformation were also a part of our experience, but with a significant difference. The period between the moments of crisis (which the Chinese understand rightly to be *both* opportunity and disaster) was so long that we tended to forget them in the mean time. Of course, we all knew about the "times of trouble," but they occurred in another time and, often, in another age. Things just went along as they always had, staying the same or gradually improving.

In the world of business, this perception finds expression in the Standard Business Curve, a graphic seemingly emblazoned on the forehead of every MBA. Things start slowly, with a bare minimum of systems and products, followed by a take-off period when systems and products (to say nothing of plant, facilities, and employees) are added at something approaching an exponential rate. Finally, growth levels off, or proceeds upward at a gentle predictable rate, following a line that hopefully projects out to infinity.

There is, however, one significant piece of data that never appears in the graphic representation of the Standard Business

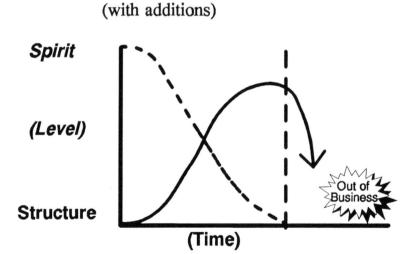

The Standard Business Curve. . . . .
(with additions)

Curve. That datum is common knowledge to every schoolboy, but somehow it escaped the attention of the organizational theorists. Simply put: "What goes up will come down." It is never a question of "if," only "when." Sooner or later, the market will change, the product will become obsolete, the competition will intensify, the financial market will fall apart. Someday, somehow, somewhere, that rising business curve will come down.

Business types might be forgiven for the oversight, however, for the "when" seemed so far in the future that it didn't seem to warrant attention. Besides it was an unpleasant thought. In those days, we thought in terms of a product life cycle that might last for 50 years, and businesses seemed to last for ever. If the notion of personal immortality, in some heavenly realm, had passed from popular favor, corporate immortality made up for it. Under the circumstances, progress and change came slowly to say the least. Indeed, there was a not-so-funny joke about "progress by funeral." Things only changed when the old folks died or retired. Until then new ideas and fresh blood simply had to wait.

THE "SPIRIT LINE"

There is one more important piece of information which never made it onto the Standard Business Curve. It is what I call the Spirit Line. Briefly stated, this line tells us that the level of

Spirit is inversely proportional to the level of structure. More directly: as Structure goes up, Spirit goes down.

I am not sure that there is any way to prove this, but I believe it is the universal experience that all organizations start out in High Spirits. If it were not so, they probably would not have started at all. Everything is possibility, the slate is fresh. Excitement and innovation abound. That is the up side.

On the down side, there is another universal experience. Confusion also abounds. While innovation is always to be cherished, the truth is that it can be taken too far, especially when the wheel is re-invented for the umpteenth time. Sooner or later, somebody says the fateful words: *We have to get organized.* With that, structures, procedures, tables of organization, and all the rest, make their appearance.

Some of the original folks never get used to the new order, but most agree that getting organized is essential. There is, however, a price. Structure constricts the free flow of Spirit. That is by no means bad, for with structure and order come those qualities dear to heart of every business person: efficiency, effectiveness, and profitability. Profit can rarely be returned when all available assets are devoted to re-inventing the wheel.

For a time, and hopefully a long time, the benefits of efficiency, effectiveness, and profitability roll in. Unfortunately, however, as that time stretches out, people tend to forget that what drives it all is the Spirit of the place. Structure is only the highway upon which Spirit moves. Naturally, there is much useful work to be done keeping the highway in repair, adding improvements and controlling the traffic. This is called management, good management.

As long as the highway runs in some useful direction, and Spirit is content to travel that way, all well and good. But when the direction of good business changes, or Spirit tires of the same old view, difficulty emerges. It is time to turn off the road. And if Spirit has gotten used to riding in a luxurious sedan, difficulty turns to trouble. Somehow limousines don't quite make it across open country.

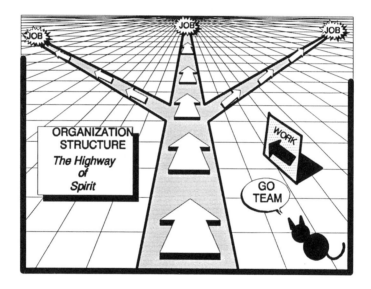

The hope, of course, is that the highway will proceed forever towards unending business opportunity. That the business curve will always go up. But as every schoolboy knows, what goes up will come down. And when it does, fat, lazy, sagging

Spirit will not be up to the challenge. Unfortunately, when the butterfly lands and chaos breaks out, the Spirit is weak.

For those who never added the "Spirit Line" to the Standard Business Curve, or who forgot the centrality of Spirit under the press of everyday affairs, the response at such a moment of ending is as understandable as it is futile. Frantic emergency actions are taken to shore up the structure and some-how get one more mile out of the old machine. Failing that (or in addition), land office business emerges in Golden Parachutes for executives and job security contracts for employes. While these various strategies may work for a while, keeping people busy and hopefully feeling better, the truth of the matter is that when it is over, it is over. Ultimately, there is nothing to be done.

More accurately, there is nothing to be done with the structural side of things. There is plenty to be done with the Spirit.

RAISING SPIRIT

If the model given to us by the chaos theorists is dis-turbing, suggesting as it does that all Open Systems eventually go to chaos, it is also hopeful. At the other end of chaos, there is the possibility of renewed existence, not along the same old lines, but actually an improved existence. There are no guaran-tees of course, but if renewal is to occur in the organization, raising Spirit is a must. When the Spirit is up once more, it then becomes possible to generate new products, profits, and structure. So how do you raise Spirit?

The subject of raising Spirit, if treated at all in the world of business, is usually handled under the heading of "motivation." In application, this turns out to be little more than the good old pep talk with some new bells and whistles, like recognition and rewards.

Pep talks are fine, but their utility diminishes when the team is not only losing, but virtually wiped off the field. As for recognitions and rewards, there is little to offer when bankruptcy, in one form or another, is the corporate reality. Pep talks under those circumstances do more harm than good. Everybody knows it is a charade, and it would be better to tell the truth and go home. There is a time for motivation, but not at the end. Something deeper is required.

The art and science of raising Spirit is not unknown. Indeed, humankind has been practicing it for all of recorded history, and undoubtedly before. It is called *griefwork*. If this term is unfamiliar, the constituent words are not. It is quite simply the work of grief, or what grief does. In most cases we tend to experience grief as something that happens to us at those moments of ending, as opposed to a process that enables us to move from one state to a new one. This is quite understandable, for grief is intensely painful, and it is more than a little difficult to look beyond the pain of the moment to see the whole process.

More than twenty years ago, a small group of people, with Elisabeth Kübler-Ross[1] in the lead, looked beyond the pain of the moment in order to see the whole. And when they did, it was discovered that there was a process of grief that moved in recognizable stages, accomplishing predictable results.

You may be wondering how we got from the relatively benign subject of raising Spirit to a discussion of grief and death. The reason is quite simple. My experience has been that significant ending, in any area of life, produces the same reactions and results for those involved. Whether we are talking about the death of a loved one, or the death of a corporation, it is all death, and the reaction is identical.

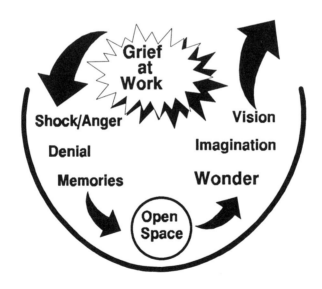

[1] Kübler-Ross, Elisabeth, <u>On Death and Dying</u>, Macmillan, 1969.

To get the point, you need only remember those times in your own experience (or in the reported experience of others) when a plant closed, a business failed, or even some element of a business (a product line) was terminated. Listen to the conversation in the hallways, or at the locked plant gate in a one industry town, and see if it doesn't sound a lot like a death in the family. And for good reason. When you have been doing a job for thirty years, and somebody tells you that it is over, you are losing more than a job. It was a way of life, and for some it was life itself. Time was defined by the beginning and end of work. Hope was articulated in terms of savings plans and bonuses. Progress was measured by promotions and company recognition. And when that is gone, who are you, and what will you become? How do you tell your kids that the college education they expected can't be afforded. How do you explain to your spouse that the retirement home you dreamed of will never happen?

The reaction comes with deep, explosive anger and shock that such a thing could happen, and more acutely, happen to you. Then there is denial, the inability, or unwillingness, to acknowledge that anything happened at all. Maybe it was just a bad dream that will vanish at the breaking of day?

Sound familiar? It is the normal, necessary, and productive process that each and every one of us goes through at significant moments of ending, when the Spirit is battered and hope is a four-letter word. This is grief working, or griefwork. There is no way to eliminate the pain, but there are innumerable ways to shorten the time and improve the odds for a successful outcome — a renewed Spirit, ready to get on with the business of living. It all begins with knowing the process, and being willing and able to facilitate the journey.

GRIEF AT WORK: THE JOURNEY OF TRANSFORMATION

Grief starts at the moment of ending or its imminent approach. The first phase is *shock and anger*. The actual expression varies with the language of the griever, but it is always some version of Ohhhhhh....Damn!!!!!

In effect it is purely a physiological response. Breathing in and breathing out. And there is a reason. With the insult of ending, people are likely to go into shock, and that often

means stopping breathing. If you notice somebody in that condition, the shoulders are typically hunched forward. This is taken to be the posture of misery, which it is, but it is also very difficult to breathe, and without breath, life stops. So the first order of business it to get the patient breathing again, and shock and anger does the job.

This phase is quite noisy, and it is not uncommon for managers and others, who may be standing about at the time, to try and calm things down with direct orders or words of consolation. However, if we understand what is going on, that response is not only unhelpful, it is counter-productive. The point is to keep the patient breathing, and if saying Oh, Damn! (or stronger) does the job, so be it.

Shock and anger is emergency first aid. Administered quickly, life continues. But as a long-term strategy, it leaves a great deal to be desired. Some people, however, get stuck at this point, and spend the balance of their natural life in shock and deep anger. You find them in the back offices of companies that have gone through some transformative moment, and these folks never quite seem to get over it. At worst, they appear paranoid and their anger erupts at unpredictable and inappropriate times. Not exactly good for business.

The sad part is that this condition is largely preventable, for had the people been encouraged to fully express their shock and anger, when such expression was appropriate, those feelings would probably not have gone underground, only to reappear later. There seems to be some finite quantity of feeling which will come out — now or later. Those who propose to care for the Spirit and facilitate the process will insure that sufficient time and place is given for the natural process to unroll.

Denial

The next phase of griefwork is *Denial*, which often appears to be a terrible waste of time. In spite of all the evidence to the contrary, those involved persist in the obvious delusion that the end has not come. Were it a plant closing, the conversation would constantly turn to the notion that, "They aren't really going to close things down. It must be some kind of a trick just to get further contract concessions. After all, we've heard it before, and it never happened." The fact that it has now happened doesn't seem to register.

All exhortations to deal with reality fall on deaf ears. The people just can't hear. And they shouldn't. Not at that moment. Denial performs an essential function rather like the bandage on a wound. It provides protection so that healing can begin. When the pain of ending is so severe, it quite literally can't be dealt with, and denial offers anesthesia. Were it possible to crack through the denial and force the folks to acknowledge reality, it is quite likely the process not only would be retarded, but actually reversed. The folks would return to shock and anger.

Once again, the role of those who would care for Spirit is to provide the time, space, and permission for the process to take its ordered course. Any effort to move for early closure, or worse skip a phase, will be entirely counter-productive.

As with shock and anger, people can also get stuck in denial. Such folks are not unknown in the corporate world, for never having acknowledged the ending of the old, they are incapable of dealing with the new, no matter how attractive that "new" may appear.

Memories

There will come a time when the reality of the situation finally can sink in, at which point the next phase, *Memories*, will begin. Superficially memories look and sound a lot like denial. But there is a difference, for now the fact of ending is acknowledged, if not totally accepted. All the memories of what happened, didn't happen, or might have happened, pass by.

The process is tedious to bystanders, for it appears that folks just keep talking about the same thing over, and over, and over again. Conversation focuses on the instant the bad  news was delivered, and the question is, where were you at that dreadful moment. Everybody remembers in precise detail with endless repetition.

What sounds like boring repetition has a pattern and a purpose. The pattern is to start at the moment of ending, and work one's way backwards through all of the events leading up to the terminal moment. "I remember I was standing by my truck just waiting to go out on delivery, when Harry came running up with the news." Then silence.

Then again, "I was standing by the truck...Harry came running up...and you know minutes before I had just loaded in that marvelous new material my customers had been waiting for."

And again, "I was standing by the truck...Harry came running...marvelous new material...and you know, last week we had our most successful week ever..." So it goes, backwards.

What seems like pointless repetition is, in fact, accomplishing some very important work. There is a purpose. With each turn, the life history of those individuals, and that group, is effectively being rewritten to take into account the new reality. It is honoring the heroes, and the heroic events, so that they can be let go. Without that acknowledgement and honor, the tendency is to hang on to what was, not consciously perhaps, but hang on nonetheless. When getting on to the future is the issue, hanging on to the past is not helpful. In a curious way, one has to go backwards in order to make progress.

There is another valuable aspect to memories, which is to take inventory of current assets. Much of what is over, is over, but there will inevitably be certain things, called experience, that will serve well down the road. Reviewing this material is an essential precondition for placing it in order, ready for the next step.

For those concerned with caring for Spirit, the advent of memories offers an occasion to actually do something. Whereas the appropriate response to shock and anger followed by denial is to let it happen and listen, with memories, some formal activities to speed the process can be initiated.

There are many names and forms possible here, and none are automatically "right," but the ancient institution of the Irish wake may be prototypical. For those who have never experienced such a thing, a short description will be in order.

Sometime after the departure of a dear brother, the clan assembles at the neighborhood pub. All through the night, and on into the wee hours, glasses are raised and tales are told. The Dear Departed is celebrated in song and story, the good parts and the bad. When it is all over, more usually with the rising of the sun, the heads may be hurting, but somehow the Spirit is refreshed. It is a new day and time to get on with life.

On occasion, I have actually held wakes for corporations and other organizations. I didn't call it that, but the intent and the effect was the same. As it turns out, you don't have to be Irish to reap the benefit.

Open Space

When the memories are over, a very solemn moment is reached: *Open Space*. Open space is quite literally what the name implies, nothingness. It is all over.

Open space is experienced initially as despair, because there is nothing left to hope in. All of the structures, procedures, relationships, that used to give life meaning are gone. There is nothing to count on, which means there is nothing to pin your hopes on. The pain of despair is intense, but it is also cathartic, carrying out the last remnants of what was. It is the final "letting go."

For some, the pain of despair can be so intense, and the fear of losing everything so profound, that the last moment of letting go never fully arrives. For them the balance of their natural lives will be lived in despair. Having nothing to look forward to, and nothing to fall back on, they become caught in the awful middle ground of meaninglessness. In the world of work, such people can be seen filling their days with trivia. Punching the time clock is all that matters.

Others, probably the majority of us, somehow use the moment of despair in a positive way. Rising above the pain, or maybe more accurately *embracing* the pain, the cathartic properties of despair are allowed to do their work. In a curious fashion, the way out of despair is not to avoid it, but to embrace it. How all of this works, I don't know, but I do know that if embraced, the pain of ending undergoes a subtle but profound transmutation. The searing agony is replaced by a bittersweet sense of peace.

Silence

With the arrival of the sense of peace comes *Silence.* Beyond shock and anger, denial, memories, and despair there is silence. There is no clock that can measure the duration of this moment, indeed it seems to be moment totally out of time. Calling this a pregnant moment seems almost trivial, but very true. Holy, if we can still use that word, seems more appropriate. This is the moment of creation. With everything gone, nothing remains to prevent the emergence of the new. It is all over. It is all potential. At this instant, if ever, differences can be perceived that truly can make a difference. The moment of silence is therefore also a learning moment of the profoundest sort. The time of High Learning, if ever High Learning is going to take place.

Wonder and Imagination

The gift of silence is received and appropriated through wonder and imagination, which together create the path through Open Space. With nothing standing in the way, all is potential. Suddenly, all those unthinkable things that could not be done "because..." are now thinkable and maybe even possible. You hear people saying, "I wonder if...." When wonder and imagination combine, something truly magical occurs. It is Vision.

Vision

    Vision is the picture of some future state that we hold in our head. But that is a very bland statement, for vision does not so much depict a new reality, as create it. We do not follow our vision. We are driven by it, even possessed. There is an element of compulsion here. What vision lacks in concrete detail (plans, budgets, designs, and the like), it gains in sheer power. Vision is Spirit bursting out in new and powerful ways.

    The power of vision is not surprising given its heritage. Born in chaos, baptized in shock, anger, and denial, vision emerges from the ashes of our memories, in silence. Then, through the alchemy of wonder and imagination, it is transmuted into a consuming passion. That is power.

Wonder plus Imagination create .....

Contrast all of the above with the current fad: Vision Statements. Typically the product of a committee seeking to instill rationale and purpose in an organization that has seemingly lost both, the Vision Statement is a pale reflection of its namesake. More often than not, the Vision Statement lacks the one thing that could make it meaningful: passion.

But then passion is messy and hardly measurable. By definition it is out of control. For managers and organizations, still living under the illusion that they really are in control, messy, non-measurable passion is the last thing they want. At the same time, there remains some intuitive sense that having a vision is a good thing. And so the Vision Statement, a superficially logical document, usually containing everything but the kitchen sink, and produced with much thought and great effort.

True vision hardly needs to be stated, and is never the product of thought and effort. It emerges despite all efforts to contain it, from a place in our consciousness that seems to have little to do with thinking. Call it irrational if you like, provided that word is used without prejudice. Vision is pre-rational or sub-rational, the very ground and foundation from which rationality emerges. Given vision, you may figure out how to implement it, but you never think your way to vision. It is always the gift of chaos, the product of transformative High Learning.

NOTES FOR CARETAKERS OF SPIRIT

There is a special role for those who would take care of the Spirit of an organization as it negotiates the process of transformation. In a curious way, there is little to be done, and much to accomplish, for the accomplishment occurs more by *being there* in sensitive ways, than by doing a whole mess of things.

To this point, we have considered the role of the caretaker of Spirit through the stages of shock/anger, denial and memories. Dealing with the first two is largely a matter of creating the time and space for people to engage in the natural process. Direct help at that point will usually only hinder progress. When it comes to memories, the role becomes somewhat more proactive, and in some cases might actually be formalized, much in the way of an Irish wake.

With Open Space, the role becomes, if anything more subtle and critical. In the midst of despair, the natural tendency is to offer hope when none is apparently there. It seems the kind thing to do. But that act of kindness may prevent precisely the most essential function of despair, the cathartic moment when letting go of everything actually happens. Knowing the process of griefwork, the caretaker is in a position to suggest that "this too will pass" but not without the pain of letting go. Not that pain is a good thing in itself, but one may be assured that painless release usually means either that the object of loss was lightly held, or that the process is incomplete. For the most part it is a matter of being there in deep empathy, which enables the process of embracing despair and provides support in the midst of genuine surrender. This is the time for a "good cry," when an arm around the shoulder or a gentle touch is all that is really required.

In the moment of silence, the natural tendency again must be resisted. It is our wont to fill up silence. In the case of organizations undergoing transformation, this usually means the latest edition of the corporate plan, or the pet projects of those on high, which were never completed. All of this is presented in the spirit of giving people something to keep them occupied when they have nothing to do. The effect upon the process is chilling, for in the event that the people actually listen (which thankfully doesn't happen often), they will be deprived of an instant of pure creation, an experience never to be forgotten. And the organization will be deprived of the people's capacity to see differences that make a difference, which is the mother lode of innovation and breakthrough. In a word, everybody loses.

There is, however, one thing to be done, which may seem rather inconsequential given the gravity of the situation. Ask a question. *What are you going to do with the rest of your life?* The reason is that questions create space as opposed to statements, which make closure. In this case we need plenty of space if the full value of the moment is to be realized. And this question in particular creates precisely the sort of space required. Raising the issue of "the rest of your life" suggests that there will be one, if desired. It is that possibility which lays the

ground work for moving out of silence into wonder and imagination. It is always helpful to fill in the blanks with small sug-
gestions about what might be done. But again it must be empha-
sized that the most helpful thing is not the answer, which all
people must find for themselves. It is the question that sets the
ball in motion.

Once the question is asked, and wonder and imagination
are engaged, vision will not be far away. When vison appears,
movement occurs and new life and new organization become
possible.

Chapter V

The Journey Continues:
The Four Stages of Organizational Life

After vision — what? Presumably we get back to work, and once more the Standard Business Curve begins to rise. New structure, new products and profits, all carried aloft on the wings of a renewed Spirit. True.

But there are also occurances along the way to the future which are not, so far as I am concerned, described by that famous curve. Maybe it just analogy or metaphor, but it seems to me that what we now know about the process of birthing suggest some useful insights regarding what happens to our organizations. The work in question was done largely by Stanislav Groff[1] as reported in his book, *Beyond the Brain*. Drawing upon physiological and psychoanalytical evidence, Groff describes what people have described as happening to them prior to birth. I leave it to Groff's colleagues to judge his findings, but I use them here, at least the basic description of

---

[1] Grof, Stanislav, <u>Beyond the Brain</u>, SUNY, 1985.

the process, only because it has been helpful and suggestive to me in making sense of the ongoing life of an organization.[2] Herewith the Four Stages of organizational life.

BLISS

Life in the renewed organization, as life in the womb, begins as bliss. There is endless space to do whatever the heart might desire. New ideas are welcome because they need not

---

[2] It is tempting to think that the relationship between life in the womb and organization is more than analogical, and speculate that the similarity exists because both are open biological systems, although on a vastly different scale. But the evidence for that is scant indeed, and the establishment of the relationship is far beyond my purpose.

contend with old ideas for attention. People, products, and procedures can proliferate without conflict. There are glorious, and seemingly boundless, open spaces in which to roam.

The possibility for innovation is without limit. At the same time those wonderful innovations to often go without implementation simply because there are not enough hands (resources) to put them in place. Truth to tell, everything might stop right here, disappearing like a good idea whose time has not come. Alternatively, the troops and resources may be assembled to get the show on the road.

TIGHT QUARTERS

As the new people and products enter the scene, the endless open space becomes a little crowded. At the beginning, the influx is comforting. It is nice to have all these new folks about, for many hands make light work. Or something like that. Shortly, however, the aphorism changes: Too many cooks spoil the broth. The pleasure of comrades on the journey is diminished when everybody is trying to do different things in the same time and space. Sooner or later, we have to get organized, and the system is born.

Separate offices, new departments, procedures for running it all and keeping it all running emerge. At best, the end result is a well oiled machine. At worst it is a bloody nuisance as people fight the system, trying to get the job done.

But new systems save the day, and it is perceived that better organization will overcome the mess. Whereas we used to just talk to each other, now we have to "communicate." The purchasing department replaces the quick trip to the hardware store for the necessary gizmo. And travel vouchers: sometimes it seems that it takes more time to compute the silly things than to take the trip and do the business. But this is called progress.

Somewhere along the line, it occurs to a few enlightened souls that more energy is actually being expended in running the business than in doing the business. Indeed there are some people who never do any business at all. They are called managers. Their function is to invent and operate the systems so that other people can do the business. But the growth continues.

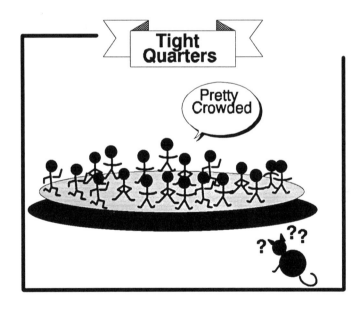

One fine day, space runs out. There is scarcely room to think, and no room to do anything useful. New thoughts and approaches run smack up against "the way it has always been done." New activities are shoehorned into odd nooks and crannies.

Meanwhile, the organizational effort proceeds. Reorganizations, rationalizations, all are aimed at keeping everything together. Systems are piled on systems, coordinating other systems, and new systems emerge to coordinate the coordination. Where does it end? Especially when it appears that the energy required to run the new systems outweighs the energy saved by their implementation? We have clearly reached the point of diminishing returns. If this were a baby, we would say there is simply too much baby and too little womb. It will get worse.

THE TOXIC SYSTEM

There comes a time in the womb when not only is there too much baby for comfort, there is too much baby for the plumbing system. The womb goes toxic. No longer is it possible to remove the natural waste products of life; life will have to radically change or cease.

As unpleasant as the toxic womb may be, there is a purpose. It sets the essential conditions for birth. Humanitarian concerns might dictate an effort to clean up the mess, but in a fascinating way, any effort in this direction will only retard the process. Birth would be delayed or not happen at all.

If we can accept Groff's description of life in the womb, it becomes clear why nobody ever would want to be born. Taking a trip down a small dark passage, to an unknown destination, is not attractive, unless the alternatives are infinitely less

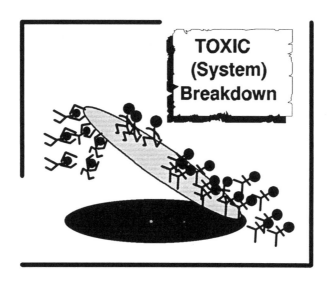

attractive. The toxic womb sets those unattractive alternatives. To stay is to die. But it is worth while noting that this is all part of the plan. The wonderful mechanism by which we come into the world includes a period so unpleasant that we literally have no choice but to be born.

I believe that the same sort of event occurs in that larger womb, our organization. As the system grows, we pass from crowded quarters to a condition in which the whole system goes toxic. For a period of time, the emergency squads go tearing about the place plugging the leaks, and trying their level best to make life livable once more. Call those squads stress reduction seminars, job enhancement initiatives, conflict resolution programs or whatever. They all have the laudable intent of detoxifying the environment, or at the very least, providing the organizational "gas masks" enabling us to remain in conditions no sane human would choose.

For a time, the detoxification effort is successful, but sooner or later, the environmental toxins will overwhelm the inhabitants. At the moment, that seems like disaster, but there is a silver lining in the cloud. The real problem is not that the system is malfunctioning, but rather that we have outgrown the system. Here, as in the womb, the efforts at detoxification are not only futile, they are counterproductive. It is time to go. The choice is clear. Evolve or die. At the very least, we must move to a new nest, having fouled the old one.

Perhaps it is totally fanciful, but when considering the larger organization of which we are all a part, Planet Earth, I wonder if the present environmental crisis is not part of a similar plan. No right-minded person can doubt that going the way we are, we will not (as a species) go very far. For the present, the environmental cleanup crews are hard at work, and in small places are making small progress. Yet the issue is not fixing the system, but creating a new way of being productive on this earth. And the problem is, doing that will require a basic change in what we hold to be valuable and useful. A change in consciousness of what it means to be human, and what a full human life is all about. Conspicuous consumption must be replaced by *appropriate* consumption. Self-worth must be measured by contribution and not acquisition.

That change in consciousness will not come easily, and is impeded more by the fact that most of the ways we know how to engineer change simply are not effective. Given a nice linear, logical problem to solve, we are superb. But in this case, the fundamental premises are different, which means that our logic is without power. Somehow we just have to take the leap into a very different way of being, and needlesstosay, nobody wants to do that. So perhaps Mother Earth is about ready to push us out of the nest. Into oblivion, if that is our choice. Or on, to flight. Wouldn't it be odd if the current ecological disaster were the most beneficial thing to have recently befallen humankind?

DOWN THE TUBES

The last act is both the end and the beginning. I apologize if the image of "down the tubes" is distasteful, but it is literally correct when speaking of the exit from the womb, and common usage when describing the demise of an organization. In short, it fits.

It is a journey nobody cares to take. It is the end of a known, and sometime comfortable, way of being in exchange for nothing, or the unknown, which prospectively may appear the same. For businesses it is Chapter 11, a fire sale, or just plain disaster. Call it the dark night of the soul, chaos, or the end, it all amounts to the same thing. The death of what was.

But with death, we now come back to the griefwork cycle. Even though the terrain to be covered as we go down the tubes may be unknown in detail, at least we know the plot. Once again we find ourselves engaged in the oldest story of humankind, the journey from birth to death, and around again.

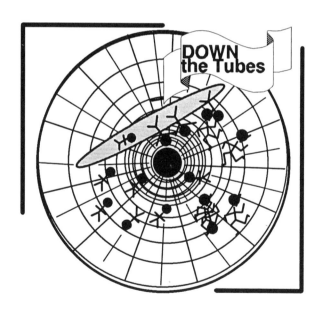

IS THIS TRIP NECESSARY?

The cycle of organizational birth and death looks rather
like the Eastern Wheel of Karma. In the typical Western understanding, this wheel describes an endless circle of life and death,
through innumerable incarnations, with no way out. That is true
only in part, for given the will and the way, the cycle can be
broken. The evolution of consciousness may proceed.

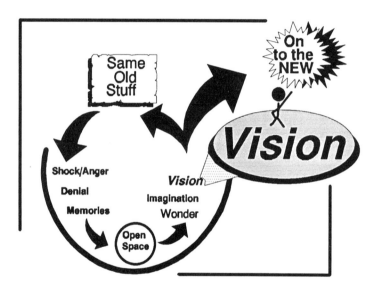

The critical step in the process is vision. And the central
question: *What is the envisioned future state?* If it is simply a
return to what was, that is the likely outcome. On the other
hand, if some higher state is held in view and the will exists to

proceed, there is at least the possibility that state will be achieved. No guarantees, of course. But the possibility exists.

The situation is not unlike that facing an executive after a major fire in a plant. Seen from most points of view, the immediate prospects are not appealing. It is a mess. But it is a mess with choices. First of all, one may accept the end as the end, board the place up and go out of business.

A second, more positive choice would be to enter the used equipment market and purchase replacements, duplicating exactly what had been lost. There is safety in this choice for one would be dealing with known quantities, and presumably the time required for a return to production would be cut to a minimum. At the same time it must be recognized that one can never quite go home again.

A third choice exists. Take the disaster as the opportunity for a quantum leap to the next generation of technology. The risk is obviously far greater, for not only would one have to deal with the trauma of the fire, but also the additional trauma of introducing the new technology. But the potential benefits in this case are obviously the highest of all.

The essential issue in all cases is vision. What do you hold as a possible future state? Add to that the necessary determination, or will, to get there. If the vision of the future is that there is no future, the choice is obvious and easy. Should the vision be "more of the same," the road to that future is fairly clear, and except for the unavoidable headache of putting things back together, the path forward is relatively easy. However, if the quantum leap forward is the vision of choice, neither the road nor the means come readily to hand. It is all new territory.

I believe a similar set of choices confronts every organization once having gone down the tubes, and passed through the griefwork cycle. When the moment of vision is reached, the issue is clearly set. Go out of business, go back to where you were, or make the leap to some new, and more effective, way of being.

Cashing in one's chips is always a possibility, and may be the only one in view. But the result is inevitably a decline in status. No longer the owner and/or manager of a plant, one goes out and gets a job working for somebody else. Given such a choice, one's self image, let's say consciousness, automatically diminishes.

Going back to "Go" (as we did in the old Monopoly game) is probably a better choice, but not without its price. Growth and learning are traded for safety, which is all well and good unless the competition has been learning and growing in the meantime. And when you do get back to Go, there may be a lingering recognition that you might have done more. The Spirit of that place can end up being tentative and risk-adverse.

Taking the high road, and going for the gold, carries the greatest risk, for as it is said, those who live on the cutting edge are apt to get sliced. If successful, however, the leap forward can put the organization at the head of the pack, and the impact on the Spirit of the place will be palpable, almost ecstatic. Even in the event of failure, there will be learning which can never be lost. Forcing mistakes to happen is not the safe way, but it certainly increases the gradient on the learning curve. In the words of the old adage, it is better to have loved and lost than never to have loved at all.

Choices made at the time of vision are important, and therefore should be informed choices. When dealing with the quality of the Spirit of a place, it is more than worthwhile to ask what the alternatives are. If we did choose the great leap forward, what would that mean? Where would we end up? The answer to that, at least my answer, will be the subject of the next chapter. In the interim we pause for some additional notes to the caretakers of Spirit.

NOTES FOR THE CARETAKERS

Those who would take special responsibility for the Spirit of an organization have much to consider, although as in the case of griefwork, the task involves being more and doing less. The appropriate role is that of the midwife. Midwives do not conceive the baby, carry the baby, bear the baby, or raise the baby, but their presence during the process of birth is very important. Their job is twofold: first, to know the way; second, to provide companionship on the journey. Occasionally, emergency action is required, but most of the time Mother Nature will take its course.

Knowing the way is not an incidental contribution. Each stage in the process tends to be so engrossing (for reasons of pleasure or pain) that the people lose sight of the fact that it is a process, and it will proceed. Following bliss comes tight quarters, and then the toxic system. In bliss, nobody wants to hear about tight quarters, and still less that the whole system will go toxic. However, the difficulty with each phase is increased if it is approached from ignorance. Knowledge does nothing to mitigate the discomfiture, but at least it doesn't come as a surprise. Further, knowing that it is just a passing phase can give much needed hope when the darkness gets pretty deep. It is the function of the caretaker to keep score, and provide comfort along the way.

Providing comfort is not to be equated with taking the burden. For in fact, nobody can do that for another. The old spiritual puts it all quite exactly, "You got to walk that lonesome for yourself. " But it sure helps to have some company along, and *being there* assumes real meaning. For busy executives, just being there may seem trivial, but I can assure you it is not. Actions may be limited to a touch or a look, but in that moment of contact comes the assurance that the journey is not made alone, a fact that can often make the difference between continuing, and throwing in the towel.

In the good old days, prior to the advent of future shock, the process that we have been describing also took place, but over such an extended time that it sometimes appeared not to take place at all. The pace has changed and shows every sign of changing even more. What used to take 50 years now can happen in one. That may see a radical, perhaps irresponsible statement, but consider the case of the introduction of PC computer chips.

Were we to graph the rate of introduction against the increase in power from one chip to the next, we would see that in this case, at least, the Tiger Ride was taking us up an exponential curve.

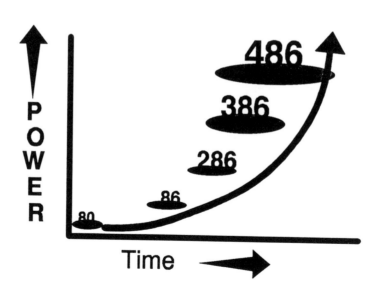

For all those in this industry, along with the rest of us who are affected by its evolution, the passage from one chip to the next generation is not just a minor improvement in technology. It comes out looking more like the introduction of a whole new business, and the consequent passage of an old one. Who would have thought that the once powerful 286 chip, heart of the Advanced Technology PC (AT), would be virtually eliminated from the market a scant four years after its introduction? The successive jumps from 86 to 286, to 386, and on to 486, each brought new possibilities for software, and new applications, all requiring new marketing, sales, and production approaches. In short, whole new businesses. Those who sought to hold on to the 286 business have found themselves holding the short end of the stick.

When described in the relatively neutral terms of technology, the rate of advance is interesting, even exciting. But seen from the human dimension, this rate of advance could become, and indeed has been, a holy terror for some of those involved. Plant closings, market downturns, layoffs, rapid obsolescence of skills. More than enough to turn your hair white, and cause nostalgic flashbacks to the good old days of job security. Some would restrain the rush of technology, but that is to hold the Tiger by the tail. Not a very attractive option.

If we are to ride the Tiger, it behooves us to ride well. That means learning to negotiate the process of birth and death in our organizations with intention and skill, lest the Spirit become severely battered. Under the circumstances the role of caretaker is not trivial. Indeed, it may be the most important role to be played, especially if Spirit is the most important thing. The bottom line so to speak.

Chapter VI

The Stages of Transformation:
Where are we?

There are stages to organizational life, very much as there
are stages to the life of an individual, and I suppose to the life of
the whole species. Identifying and naming those stages is a
useful way of keeping track of where we are, and projecting
where we are likely to go. But it is always important to remember that the stages are arbitrary and depend in large part on what
you want to keep track of. The essential criteria are do they
work and are you able to see, and understand, in a new, deeper
way? Rarely, if ever, is it a question of right or wrong, but rather
utility.

For our purposes, the central "what" to be tracked is the
Spirit of a place as it evolves and manifests itself in new forms,
for that is what transformation is all about. The forms themselves
are like the crab's shell, good for a period of time, then dis-
carded, but never to be confused with the crab.

Actually, a better image for the tranformational journey
of Spirit is the butterfly. As we all know, butterflies start out as
caterpillars, then go into a cocoon (chrysalis), only to emerge
with beautiful colors, and wings to fly. What we may not know,

or always remember, is what happens in the cocoon. Inside, far from curious eyes, an incredible process takes place. The caterpillar quite literally dissolves into a sort of primal ooze, a real mess. And then, following directions coded somewhere in that mess, the raw protoplasm re-aggregates, and what was once a crawling beastie now has the hardware to fly.

Put somewhat differently, and probably not in accord with the best biological thinking, the caterpillar goes to its essence (we might say Spirit) and then reforms, or better, transforms, to a totally new way of being. It is  a journey from form, to nothingness, to new form. And there is no way of getting from caterpillar to butterfly except by passing through the void.

In the previous two chapters we followed the Spirit of a place as it moved from one way of being to a new one. In and out of chrysalis, so to speak, through the cycle of birth and death. It is now our purpose to consider the several forms that Spirit assumes along the way, laying the groundwork for suggesting where we are, and where we might be headed.

Describing the forms of Spirit, particularly the Spirit of an organization may seem a rather new undertaking. But conversations about Spirit have been going on forever, or so it seems. And some of them have gotten fairly detailed and precise. Most of these conversations have revolved around the Spirit of the individual, which in the East is often referred to as consciousness. The Eastern schema for describing the evolution of consciousness are multiple and complex, but basically seem to break down into seven steps, or multiples thereof. The first and the last steps are essentially nothing or "void," about which not much can be said. However, the intervening five stages may be described very briefly as follows: *Body, Mind, Intellect, Soul, and Spirit*[1]

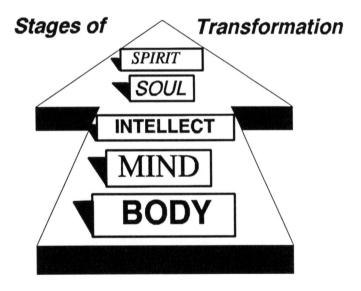

Stages of **Transformation**

SPIRIT

SOUL

INTELLECT

MIND

BODY

---

[1] I owe this nomenclature to Ken Wilbur who has made a lifetime study of the subject. He has written extensively, but a handy reference to this particular listing may be found in his book, <u>Up from Eden</u>, Anchor Press, 1981.

As *Body*, Spirit is quite concentrated on the here and now. Physical concerns are primary, which is not all bad, but definitely has its limitations.

These limitations are thrown off in part, as Spirit evolves to *Mind*. The body, of course, is still very much there, but now transcendable in some ways. A major new element is language, which enables us not only to experience the world, but to talk about that experience, share it, and reflect upon it. Thinking, in short, comes into play.

The next level, *Intellect,* includes the preceding ones, but adds the element of awareness, or consciousness of our consciousness. Not only do we have experience (body) and think about our experience (Mind), but we are *aware of experiencing the experience.* In practical terms, we begin to exist in a rational world which really doesn't exist anywhere except in our minds. This means that we can improve the quality of the activity of our minds and bodies by measuring our present state against some imagined future state. Life planning becomes a possibility. But it also means that we often confuse the constructs of our minds with reality itself, and come to the mistaken conclusion that what exists is "I." In a word, the ego is born, which as we all know, is subject to being egotistical, or stuck on itself.

Breaking through the ego to rejoin the rest of the world is accomplished when we reach the level of *Soul*. My use of the word soul is not unlike Black American street usage, as in the phrases, "He/she has got soul," or "Soul Brother/Sister." Having soul is getting it all together and really being related. Instead of being "stuck in one's head," all elements are integrated: body, mind, and intellect. By the same token, the integration extends to the world at large. No longer the island ego, life can now be lived in genuine community.

It is noteworthy that each form of Spirit includes, and requires, the preceding ones. Obviously you can't have a mind without a body, and the Intellect (even though it sometimes tries to get along by itself) requires both. When Spirit begins to operate at the level of Soul, which represents a real passage beyond ego, the ego is still there and still necessary. After all, you can't transcend something you never had.

The last stop along the journey of Spirit is *Spirit*. I recognize this is a tautology, but how else can you say it? Spirit has become fully itself. Not that the body, mind, intellect, and soul have all disappeared, but rather, Spirit is no longer stuck with any of them. More accurately, Spirit can be any one, still be itself, and always know that it is more than all of its prior manifestations. Thus we can expect inspired performance from the body, enthusiastic mental activity, and spirited rational analysis. Most of all, Spirit can now be itself.

It is a little hard to say exactly what Spirit being itself might look like, for by definition we have passed beyond most of the forms at which we might look. Yet I think we all have intimations of pure Spirit, which occur at those moments of outstanding human performance that just flow. At such times, considerations of time, space, and effort do not enter the conversation.

Obviously we do not have to go to the East to find a suitable developmental schema. The choice to go eastward is in part to acknowledge a substantial body of thinking which never quite surfaces in current organizational theory. There is also a deeper reason. Somehow most Western schemes stop at precisely the point where things really get interesting. Ego (intellect) appears to be the end of the line, and all we have to do is create bigger and better egos. But that, I think, is precisely the problem. As long as the world begins and ends with me, and my ego, the possibility that we will ever create the sort of planetary community necessary for our survival is limited.

THE ORGANIZATIONAL ANALOGUE

Every organization has a Spirit which is more than the sum total of the individual Spirits manifest in its members. I certainly cannot prove this in a way acceptable to the canons of science, but I don't think that is necessary. Walk in the door of any establishment, and you can almost smell it. It can be happy or sad, energetic or lethargic, but you know it when you met it. And even in its absence, you know that it is there. We say, "That place just ran out of Spirit."

Although we don't ordinarily think about it, the Spirit of a place not only changes with the hour and day, it also evolves over time. The stages of evolution would be only a matter of academic curiosity were not the quality and power of the Spirit of a place very directly related to the bottom line, no matter how that bottom line might be measured.

The following developmental sequence is nothing more than the organizational analogue to the traditional evolutionary stages of consciousness described above. The terms are not original, having been borrowed from the work of Frank Burns,

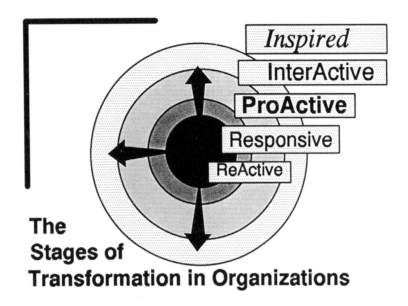

The Stages of Transformation in Organizations

Linda Nelson and Russell Ackoff. However, I believe they are used in an original way which may occasion some difficulty. If confusion arises, I ask only that you allow the terms to function in context and not worry about whether my usage is "right." After all, it is only a story, but it happens to be my version.

94

ReActive

The first manifestation of Spirit in an organization is ReActive. That may sound quite negative, but I don't mean it that way. At its best, the ReActive Organization is electric with energy. It is primal, physical, vital, alive, and just about as aggressive as it ever will get. Fancy offices and elaborate procedures are not to be found, in part because of the youth of the organization, but it is also a matter of style. These are the days "out in the garage," where frills are not tolerated and anything unnecessary to getting the job done is either not noticed, or viewed with disdain. One client lived out its days as a ReActive Organization over a Chinese laundry. Space was purely functional, and when new projects required a different configuration, walls were simply uprooted from the floor and reset, leaving behind the marks of their previous position. When I asked why they didn't clean up the mess, it became obvious they never had noticed and didn't have the time. In addition, they rather liked the marks on the floor as a reminder of where they had been.

The central character is the entrepreneur, the driven individual who seemingly brooks no interference with the realization of his or her creation. Not knowing, or not caring, why things have always been done in a certain way, this person seizes any viable possibility for the implementation of the "good idea." And any threat, perceived or real, to its implementation elicits instantaneous response. The personality and drive of the entrepreneur absolutely set the tone, establish the Spirit.

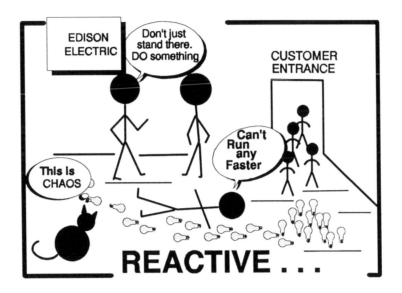

Life in a ReActive organization is truly exciting, and should the organization survive, this period inevitably becomes "the old days." But survival is by no means guaranteed, and life in constant turmoil, self-generated or externally initiated, can become very trying. For some people, this is a chosen way of life, and when things calm down, if they ever do, those people will move on to some new garage or Chinese laundry. Alternatively they will stay and live in the world of the old days.

When the basic mode of action is always ReAction, the range of options is limited, and the amount of energy required just to stay in the same place is enormous. The strength of the organization eventually becomes it Achilles heel. Constant innovation turns into re-inventing the wheel. Life without constraint leads to endless firefighting. And reacting to each challenge, as if it had never happened before, requires a level of energy and dedication almost impossible to sustain.

The cost is paid in terms of employee burnout and a decline in customer satisfaction. Both of which are inter-related. It is difficult, if not impossible, to satisfy customers with exhausted employes. And customers, once attracted by the energy of the organization, and the innovation of the product, find that attraction diminished when the product is never delivered on time because all the organizational energy is devoted to putting out brush fires.

Sooner or later, chaos strikes. Not just the old, everyday, self-generated disorder, characteristic of the ReActive Organization, but real chaos, tripped off by some trivial event (a butterfly landing), but generated ultimately from an inappropriate fit with the environment. It is time to move on. But to what?

The specifications for a useful future state are fairly clear. Needed is everything previously held, but done in a way that is responsible to the needs of the customer (external environment) and the employee (internal environment). Good product, delivered in good time, without killing everybody just to get it out the door. It sounds easy and achievable, all we have to do is do it.

But not so fast. There is a price tag. All those folks, who discovered their purpose in life through their ability to put out fires will now become responsible for insuring that fires don't happen, at least as often as they used to. And those other folks, who experienced their full personal power in pushing the boundaries and constantly inventing the new, must now be satisfied with doing the same thing over, and over again, at least until there is a modicum of efficiency and effectiveness, to say nothing of profit. A whole way of life, and for some life itself, must change. Some significant number will choose not to make the trip, and those who do will discover that the emergent Spirit charges a high entrance fee. Let go of the old in order to discover the new. With a new Spirit, there can be a new organization, but no guarantees.

Responsive

Enter the Responsive Organization. A genuinely nice place to work, and a nice place to do business. Walk in the door, and you can feel the Spirit. It is pleasant. Gone are the piles of junk from abandoned experiments, odd pieces of product and material stacked at random. In its place there is order, perhaps not elegant but functional. Shelves and cabinets contain what they are supposed to contain, available for inspection.

And the employees have a smile. Not all the time, mind you, but more often than not. The harried firefighter of old is replaced by the courteous salesperson. Work is measured not by levels of exhaustion, but by regular time periods, eight hours a day, and five days a week, holidays and vacations included.

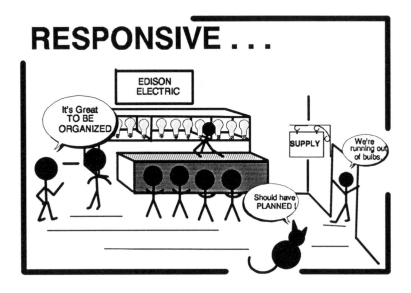

RESPONSIVE . . .

Good, responsive organizations are found all over the world. A typical manifestation is the mom and pop store. Set in the neighborhood, knowing the neighborhood, and serving the neighborhood. In the United States, one might think of Radio Shack, or in the old days, Sears. The technology level is never high, but it always works, and if it doesn't, bring it back for a no-questions refund or replacement. The English version is the local pub, a place where generations have gathered to enjoy their favorite refreshment and each other. Nothing fancy, just good folks, good spirits (of all kinds), and a little something to eat if you are so inclined. Just like home, and maybe better. The West African variety, particularly in rural areas, is the Lebanese store (in East Africa, the same institution is run by Indians). Part trading post, part community gathering place, the Lebanese Store brings east and west, black and white together for trade and communication.

The Responsive Organization is a comfortable place and sensitive to the people involved: those served, and those who serve. The Responsive Organization can also be terminally boring, just plain dull. Everything is as it always was, which is why people come, and keep coming back. The total focus is on the immediate neighborhood. The world beyond is somehow un-related. Products arrive, are unpacked, stacked on shelves, and distributed. Nobody seems to know, or particularly care, where they came from, how they were made, or indeed, if there are any new ones. Global understanding is not a strong point in the Responsive Organization, and the basic operating premise is, Don't fix it if it ain't broke. After all, if it was good enough for our fathers, it is good enough for us — a truth that seems to remain true for very long periods of time. In the interim, it is good to be comfortable. Comfort is the staple strength of the organization.

But the time arrives, even as it did in the ReActive Organization, when the organizational strength becomes its Achilles heel. With a lack of understanding, to say nothing of curiosity, combined with a very narrow focus, it is all too easy for the wider world to change un-noticed. Events, which could have been prepared for with a little foresight, roll on until they roll over the good, old, comfortable Responsive Organization. Once more it is a time of chaos, and some important choices must be made. For a small time, the firefighters, left over from an earlier incarnation, do their job. And the responsible members attempt to be more responsive to the world as they knew it, but unfortunately the world has changed.

Specifications for the desired future state, if the future is to be different from the past, are not hard to make. It would be good to retain the energy of the ReActive phase, combined with the service of the Responsive. But more is needed: rational, critical reflection, capable of sensing the winds of change and creating strategies to realize the emerging opportunities and minimize the potential threats. In short, planning, and with planning, the possibility of controlling the organization and the environment.

It sounds like a wonderful idea, waiting only to be done. But hold on for a moment. Here too there is a price tag. For all those people who found comfort in the daily sameness, and security in the assurance that things will be as they always were, the possibility of critical self-analysis hardly exists. And planning appears as a profound violation of everything meaningful in life, for if everything is always the same, why bother to plan? And if it is not the same, why bother with it at all? The process of transformation has begun once again. Some will make the trip. Some won't. But the Spirit of the place must find a new form.

The ProActive Organization is a very different beast, characterized by such things as rationality, planning, control, and power. Whoever has the power, controls the destiny of all through rational planning. That, at least, is the story. Data is collected, analyzed, assessed, and acted upon. Everything happens according to the plan.

At best, the ProActive Organization is truly impressive. A well oiled machine, self-contained, self-confident, and powerful. Until fairly recently, to be ProActive was to be among the elite of world organizations, and to such organizations the elite (or would-be elite) flocked in droves. This is the home of the MBA. Doing it all by the numbers, the MBA and fellow travelers created the megamonsters of the business world. Systems begat systems to control the flow of goods and services to the far corners of the globe. And perhaps more importantly, to control the flow of profits which returned.

Conceived as closed systems operating in a clockwork universe, ProActive Organizations created enormous wealth, power, and prestige. But they too have an Achilles heel. The source of their power is also the cause of their weakness. Their obsession with control renders them ultimately incongruent with the environment.

The essential problem is that everything turns inward. To insure control, one attempts to close every aspect of the system. No leakage, nothing operating outside of official channels. All

ProActive . . .

actions must be regimented according to The Plan. And the job of the manager is, of course, to make the plan, manage to the plan, and meet the plan.

To a point, all of this works, and the obvious success of the ProActive Organization is a measure of that workability. But that point is passed when the desire for controlling a closed system translates into a belief, indeed firm certainty, that the system really is closed, and control, as desired, is possible. This is delusion, and the price of delusion is failure.

The problem goes deeper. In an effort to close the system (or maintain the belief that the system really is closed) attention fixates on the system. The focal point is the system itself, maintaining the system, building the system, fixing the system.

For a period of time, this focus on the system works, and indeed is beneficial *so long as the external environment remains relatively stable*. That stability existed for the major part of the post-World War II years. However, when the environment radically alters, as it is currently doing, and chaos becomes part of everyday life, as opposed to an occasional occurrence, the illusion of the closed system becomes apparent, and those who have turned inward are ill-prepared to deal with the changing environment.

The problem is not that somebody is doing something wrong, but rather that they are doing the wrong thing. Those who have grown up within the ProActive Organization find themselves in the curious position of doing everything right, as they would understand "right," only to discover that the situation is further deteriorating. The reaction, understandably, is a certain desperation, cloaked by such phrases as "counter-intuitive," for it seems that things aren't happening as they were supposed to. The truth of the matter is that the current happenings in the world of the ProActive Organization are not counter-intuitive, except from the point of view of the ProActive organization. They are exactly what one would predict in the world of open systems operating in the dance between chaos and order.

It is precisely here that we now stand. At the end of the ProActive Organization, and not yet on to the next act. For the past little while, we have been attempting all kinds of "fixes." Almost any kind would do. For a period the fad was entrepreneurship (or intrapreneurship), entered into with the hope that we could recapture the vitality of our youth, increase the level of innovation, and get the well oiled machine moving again. But the truth of the matter is that entrepreneurship and the ReActive Organization are a reversion to an earlier state. It is exciting and creative, but it also means starting the process all over again from ReActive to Responsive, and back to ProActive. We have been there before. While a good shot of entrepreneurial zeal is always invigorating, it certainly doesn't constitute a permanent fix.

A second alternative is to try doing better what we think we have done well. If the problem is chaos, obviously greater control is called for. Fiscal controls, quality controls, production controls, and then controls for the controls. Outside of providing a lot of people with jobs (checkers to check the checkers), I rather suspect that the net effect has been minimal. The problem is not that we are losing control, but rather that we never truly had it in the first place. In open systems, control as we envisioned it is illusory, and in seeking greater control of the sort we thought we had, we can only fail. Which is precisely what we are doing.

Rather than moving backwards to the primal world of the entrepreneur, or digging in to secure our place, I believe we must vision forward to a rather different way of being in organization. I also believe that we have no choice. The emerging environment will push us forward, or out of the way altogether.

Specifications for a desired future state, if the future is to be different from the present and immediate past, are not difficult to prescribe. Spirit must assume the form of an open system, operating amongst other open systems, all within the largest

open system, the cosmos itself. I call this the InterActive Organization. No longer turning inward to fix the system, the focus of Spirit must also turn outward to embrace the environment. And that embrace cannot be one of hostility and fear, protecting what is "mine" and "ours," as against "yours" and "theirs." For the simple truth of an open system is that we are all in it together. It is a world of leaky boundaries and interconnections, where distinctions are noted usually only in their dissolution.

It is also a world where chaos is no stranger, but rather the constant precursor of new order. Even as the environment must be embraced, so must chaos. It is the fecund ground of new creation.

Under the circumstances, eternal structures and unchanging organizations will be a thing of the past. The *absence* of change will become the major worry, for it will mean the end of life. In this context, control will be a sometime thing. Here today and gone tomorrow. Nor will it be relinquished with anxiety and pain, but rather something approaching joy. To embrace chaos is to lose control, and that is the precondition of birth and new growth.

Strange new world indeed. But if it happened, what would it be like? Amongst other things, I think we would find that the environmental crisis turned from disaster into enormous opportunity. As we addressed the wounds of our planet we would discover new ways of being human that went vastly beyond the confines of a nine-to-five world with (illusory) job security, performing tasks that have lost their meaning. At a deeper level, we would know that wounding the planet was wounding ourselves. Not just wrong, it is crazy.

On a more personal level, we would discover that in a world constantly in flux between chaos and order, the possibility for innovation and personal fulfillment is unending. No longer blocked by the way things were, we might playfully create the new. In a curious way work becomes play, and play becomes work. Not just a reversal of roles, but a blending of effect. Call it High Learning.

And for businesses, I think it would be wonderful. The day of the limited, finite market would be over, for in open systems, there are no firm boundaries. Fighting for market share would become a ridiculous occupation, for a percentage of infinity is still infinity. Expansion and growth are limited only by our perceptions.

Sound good? Actually, I rather expect that it sounds like Pollyannaish chatter, suitable for nursery rhymes and fairy tales, but scarcely possible in the world as we know it. And that is just the point. Moving on, while attempting to maintain our present conception of the world and ourselves, is impossible. A change in conception, consciousness if you like, is essential.

Consider the alternatives. Mikhail Gorbachev has said directly what we all know. "We are already in a state of chaos." It is no longer a question of preventing something, but living with it, and hopefully through it. Were it only a matter of political chaos, we might hope someday to set the situation to rights, and get back to doing business the way we always have. A little money here, and little money there, and soon the under-lying economic factors might stabilize to the point that an or-dered political solution would be possible. Perhaps.

But there is the added dimension of the environmental "problems." A nice safe way for talking about approaching ecological catastrophe. The problem in Eastern Europe and the

Soviet Union is not so much that they are bankrupt financially (which is true), but rather that area is an ecological cesspool. Furthermore, to increase business and industry in that area, following practices they have used to date, would only make the cesspool bigger and deeper. Even if they were to use the technology available in the United States and Western Europe, the problem would only be delayed, but not solved. After all, acid rain, global warming, and toxic waste all have their major contributors here in the West. I think the evidence is coming in pretty strongly. The deep structure has changed, and the house rules must follow suit, or there won't be a house.

Adding insult to injury, we have the Electronic Connection. We all know everything (or at least more than we can digest), all the time, instantaneously. And just about the time that we think we can relax and take a deep breath, some new notice is posted on the global community bulletin board. If the physical facts of the matter don't get us, the psychological stress will. Actually it is probably a horse race. In any event, we are running just about as fast as we can run. It is time to run smarter. We will evolve or die.

Bluntly stated, the womb has gone toxic, and we are about to take the trip down the tubes. Under the circumstances, a change in consciousness is a fairly minor consideration. The consciousness of who and what we are will inevitably change, with or without our permission. It may end up being un-consciousness, but the question of change is not a question.

Actually, I think the case might be made that consciousness has already changed (for the better), and we are just trying to catch up with ourselves. In support of this apparently outrageous statement, I would point to what I take to be a fact. Everybody knows at some deep, albeit unmentionable, level that the deep structures of our planet have shifted, and consequently, the way we have played the game to the moment is no longer viable. It is not a question of knowledge, but rather the appropriation of that knowledge, and action thereupon.

The situation parallels that which pertained towards the end of the anti-nuclear movement, and prior to the outbreak of disarmament. Nobody had a doubt that we possessed the collective capability of frying ourselves millions of times over. The madness of MAD (mutually assured destruction) was apparent. The consciousness had changed, and it remained only to work out the details. Certainly, a number of details remain, but we are working on it.

If there is any truth to the notion that consciousness has already changed, and therefore we are now trying to get used to the new reality, it would then be the case that we have already gone down the tubes in terms of our prior consciousness, and have entered into a moment of significant transformation. In a word, griefwork has begun.

Taking this as a working hypothesis, we might then notice that most, if not all, of what is going on around the planet

110

is exactly what we would expect. Obviously, there is an enormous amount of shock and anger. But at least the patient is breathing. And denial is breaking out all over. The conventional wisdom is that things haven't really changed, nor are they ever likely to. But secretly, we know differently. There are even some signs that memories have begun with the emergence nostalgia clubs, formal and informal, that seem to be springing up in many of our organizations, to celebrate the days gone by. We haven't gotten into Open Space as yet, but the leading edges of despair may be coming into view.

Obviously, this is only a likely story, but if there is any truth in it, a hopeful one. If the journey has begun, several things are true. First, there is no guarantee that we will complete the course successfully. Secondly, there are a number of quite practical things we can do to increase the odds in our favor. The question, or better — *Quest* — remains. How do you get from here to there? I believe we have the makings of an answer, but the story is not over.

Looking at that answer (if such it be) is a task reserved for Chapters VII and VIII. In the interim, let me complete the description of the stages of transformation. Impatient souls will be forgiven should they choose to skip along, but before doing so, consider the fact that it may be useful to explore the future, two mountains hence. I do not believe that the present moment is the last moment. Nor do I believe that, should we pass through this moment successfully, we will have reached the end of the trail. The end is still ahead and best of all. The quest continues.

Inspired

If the InterActive organization is anything more than a figment of my imagination, and anywhere nearly as attractive as I propose, why on earth would we wish to move beyond? The simple, and perhaps simple-minded, answer: Because there is a "beyond," or at least there might be. One thing all of us apparently share is a profound wanderlust. If there is another hill, sooner or later, somebody will look over it. Past a flat earth, to a round one, then to the moon, and always beyond. Then down, into the interstices of life, past the atomic barrier, where matter ends, and pure energy begins. Into the strange world of neutrinos and quarks. Surely, if there is something beyond the InterActive Organization, we will take the trip. Or someone will.

But there will be a reason for that trip, for the Inter-Active Organization has its limitations as well. Someday we will experience it as the crab does its old shell: too small for us. Too small for our Spirit. If the InterActive Organization is characterized by the playful creation of new forms, taking joy in the fabrication of wonderful new ways to be, it is still locked into the world of form. And one day Spirit will move beyond form. It has to, if only to be fully itself.

What that will look like, and when it might occur I cannot say. But already, I think we have vague intimations of life at that level. Fleeting for sure, and always suspect in the cold light of day, nevertheless they pop up to surprise us. Perhaps we find ourselves at a concert, and suddenly, for no apparent reason, the hard seats seem to evaporate, conductor and musicians virtually disappear from our awareness. The fact that the music is Bach, Beethoven, or something contemporary, makes little difference. The difference that makes a difference goes beyond all of that. Quite literally, out of time, and out of space, we experience pure sound that makes its own time and space, and seemingly requires neither. It just is.

INSPIRED. . .

It doesn't happen often these days, but when it does NO ONE forgets. Time goes uncounted. Space does not bind.

Things just FLY

One doesn't have to go to a concert for the experience. Every now and again in world class athletics, a whole team will simply transcend good technique and hard work. Game plans mean nothing. The opposing players no longer form the opposition, but fold into the dance of superior performance. What happened? It is hard to say. It just is, was, and always remains embedded in the memory as a recollection of something very special.

I think even the drab world of business and commerce knows such moments. Of course, we try hard to stay grounded in the earthy realities of dollars and cents, bits and bytes, but in spite of our best efforts to remain where the rubber hits the road, something happens. It may start with a mundane work group, and a task nobody particularly wanted. But somehow the enthusiasm builds and the task is no longer a task due by a certain deadline. It becomes a way of life that wipes the time clock from the wall and the sense of spatial constriction imposed by the walls the office. Imagination soars and things get done that nobody could expect, or even have named before they occurred. In retrospect, what sticks in the mind was not the concrete accomplishment, impressive as that might have been, but the heady sense of free flight, when the Spirit really took off. Of course you would have had to be there to know what I am talking about, and even then, you wouldn't be sure.

Were we concerned about such things, it would be easily possible to find parallel experiences in the writings of the mystics of all ages. And if parallels offer confirmation, we might

114

suspect that the epiphenomenon or our own experience was more than a fantasm.

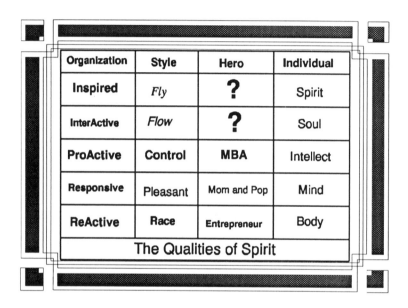

Organization	Style	Hero	Individual
Inspired	*Fly*	**?**	Spirit
InterActive	*Flow*	**?**	Soul
ProActive	**Control**	**MBA**	Intellect
Responsive	Pleasant	Mom and Pop	Mind
ReActive	**Race**	**Entrepreneur**	Body
The Qualities of Spirit			

Chapter VII

The InterActive
Learning Organization

Assume that the journey has begun. We are Riding the Tiger, and maybe, just maybe, we know where we are going. The name of our destination: the InterActive Learning Organization. The central questions: What would it look like? And, how do we get there?

What follows is, to a degree, pure vision. It may not exist, now or ever. However, this vision has been born in wonder, by way of imagination, at the incredible accomplishments possible when Spirit embraces chaos, and joyfully creates new order. Like all visions, it will be rather short on details and plans for implementation. If you expect proof of its possibility or actuality, it would be well to stop reading now. The validity of a vision, as beauty itself, lies in the eyes of the beholder. All I can do is share the vision, tell the story.

In a sense there is nothing new here, for we have done it all before. But to the moment, these accomplishments have appeared as aberrancies, exceptions to the rule. All done in spite of the "system," in spite of our best efforts to the contrary. As the young laboratory director from Dupont remarked, perhaps it is now time to do intentionally what we have been doing all along. Thus an important clue to the future are the mistakes of the past. Those funny little (or very un-funny large) mistakes that created differences that made a difference. The key to the appropriation of this future will be a change in intention, a shift in the consciousness of who and what we are.

LEARNING IS THE KEY

You may have noted that the word *learning* has inserted itself into the title. The reason is simple. The key, and characteristic, activity of the Interactive Organization is learning. Not some of the time, but every instant. Not in spite of itself, but with clear intention. The content of this learning is not (in the first instance) the facts, figures, and formulas necessary for productive life. And the format for learning is rarely, if ever, the class, seminar, or training session. Perhaps most critically, the fundamental learning taking place is not the product of an organization of learners, but rather a learning organization. It is whole, organic, and the totality is infinitely more than the sum of the parts.

So where do we start? Probably with a story. Several years ago in Mexico, I met a new friend. His name was Mahesh, and he came from India where he was vice president for human resources with a large hotel chain known as the Taj Group. We had both been invited to make presentations at a conference. His was on learning and mine, the transformation of organizations. By the luck of the draw, he went first, and I sat and listened. To my astonishment, what he had to say about learning exactly tracked what I was going to say about transformation in large systems. Not that they fell in sequence, learning first, followed by transformation. They were the same thing.

The core of Mahesh's presentation came from a study he had been doing for a number of years, in which he asked a large and various group of people to recall their deepest learning experience. Not just when they first learned to read, or finally conquered differential equations, but something, or sometime, that really made a difference. With that moment in mind, Mahesh probed further, and asked them to remember how they felt going into that moment, in the middle of it, and afterward.

With variation in detail, the basic answer was always the same. Going in, the feelings were ones of anxiety, uncertainty, discomfiture, dread, and fear, all mixed up with excitement and anticipation. In the middle of the moment, the feelings defied description except in terms of chaos, pain, and eventually no feeling at all — nothingness. On the far side of the moment, the feelings changed to exhilaration, celebration, and a soaring experience of liberation, culminating in a sense that a real difference had been realized. Something new had happened.

Mahesh had done nothing more or less than describe the feelings experienced during the process of transformation as grief worked. My conclusion, which I have been trying to understand ever since, was that learning was transformation and transformation learning. Different ways of talking about the same thing: the passage of Spirit to new form.

The story continued two years later when I, along with Mahesh, and support from the Taj Group and Procter and Gamble, convened two working conferences around the theme, "The Business of Business Is Learning." We didn't know enough at the time to call the object of our concern the Learning Organization, but that was where we were headed, and where we arrived. But there were some surprises along the way.

The first conference was held in India, Goa to be precise, and attended by a group of 50 senior executives, theorists, and consultants. The majority were from India, with a sprinkling from around the world. We used Open Space Technology (details in Chapter VIII), which enabled the group to create and self-manage its own agenda for the five-day affair, knowing only in advance who was coming, the theme, when it started, ended, and the site.

The first day of the conference was formal to say the least, as individual members of this august body presented their favorite learning theories. At the end of the day, one of our number addressed the group as follows. "I have learned nothing. While I do not mean to be insulting, I have heard nothing new, and if we can't get past what we already know, we should probably stop right now." Given the size of the egos present, a massive explosion might have been anticipated. And it probably would have occurred, except for one thing. Everybody agreed.

In Open Space Technology, there is no conference management group, design team, or staff. This means that responsibility for the conduct of affairs lies completely with the participants. It also meant that the non-existent design team did not have to withdraw to lick its wounds. We all just went to dinner.

The next day a surprising thing happened. Four previously scheduled work groups, each meeting around a different subject, suddenly found themselves talking

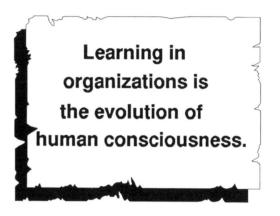

Learning in organizations is the evolution of human consciousness.

about the same thing. *Learning in organizations is nothing more, nor less, than the evolution of human consciousness.* And the role of the organization is to facilitate that process. Perhaps it was the fact that we were in India, where strange happenings seem a little more commonplace, but the truth was that nobody would have predicted the turn of events. Can you imagine an executive of an American steel mill seriously talking about the evolution of human consciousness as part of the corporate concern? The likelihood is no greater in India where executives are, if anything, more westernized than westerners, at least that is my experience. Nevertheless, that is what we were talking about — very seriously.

The conference ended in an even more remarkable way. On the final day, as we were contemplating where we had been and what might lie ahead, a young woman, the managing editor of *Business India* (the rough equivalent of *Fortune*) said, "I came to this conference thinking I would probably get a vacation (we were located at a plush resort), and maybe get a story. Now at the end I have to say that I had no vacation, never having worked harder in my life. I have an incredible story, and it has changed my life. We came to talk about learning organizations, and in spite of ourselves, we have become one."

LEARNING IS THE EVOLUTION OF HUMAN CONSCIOUSNESS

The notion that learning is the evolution of human consciousness, and further, that the fundamental function of the organization is facilitating the process, may appear outlandish at best. Some organizations, possibly — but businesses? The very thought leads to images of all members sitting in lotus position, while the competition runs off with the bacon, to say nothing of market share. Hardly practical.

However, if we leave aside the question of method (the lotus position is not the only way), I believe there is a hard core of practicality here, with a direct relation to the bottom line. Let us say, for example, that your business is stocks and bonds. The place of business, at least the center of business, is the stock market. Everybody understands that business starts with the opening of the exchange, and closes with the final bell. In between, there is all sorts of running about, buying, and selling.

The basic time/space configuration of that business is clear, and over the years your folks have created lives that match. To be on the floor for the opening bell requires leaving home at a certain time. During the day, everybody knows where they are — "on the floor." Spouses and children feel comfortable because even though they may not understand exactly what is going on, they do know that the employee had gone to the market. Then at the end of the day, with the final bell, everybody goes home, and dinner may be served.

One day, a funny thing happens on the way to the future. Actually, it was the 19th of October 1987. Apart from the fact that the market went to hell, there was an even deeper disturbance. The market essentially disappeared in physical form, only to be remanifested as the Great Computer Conference in the sky, and time ceased to be measured by the opening and closing bell. Instead the market began to run on global time, 24 hours a day. Of course, it continued to touch down at various locations around the planet, but the real action never ceased. And anyone, anywhere (with the proper credentials, a PC, modem, and phone line) could join the market. So if one were to ask, where is the market, the obvious and truthful answer is anywhere, anytime you want it.

This transformation of reality, and those words are truly appropriate, had major impact. In the United Kingdom, for example, they closed the stock market. I don't mean just for a day, but permanently. The great hall now stands empty. And the impact on all those who had previously defined themselves by the opening and closing bells at that particular place, was (is) enormous.

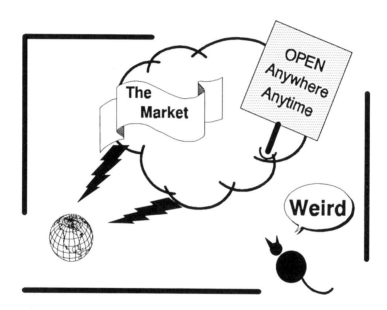

For our purposes, the critical change is in self-understanding, or to use the more appropriate word, consciousness. Were one to continue operating in the new reality with the old consciousness, one would be out of business. What used to be (or so it seemed on the surface) a fairly cut-and-dried, controllable business, firmly located at a specific time and place, had transformed. In its place there appeared a completely interactive, global enterprise, operating beyond the time and space of any particular piece of the planet. In order to stay in business, consciousness had to change, and the continuing story in the brokerage houses of the United Kingdom is how to facilitate that process.

October the 19th was a significant learning moment, a moment of transformation, a moment for shifting consciousness. Those who have managed the process are still in business. Those who haven't presumably have found alternative employment. And the process continues. From this side of the Atlantic, things look relatively peaceful as the folks settle down in their electronic cottages, dividing the global day into regular shifts. But closer to the action it is apparent that the griefwork, initiated by the transformative event, continues. Shock and anger are still around, and some folks continue in denial, firmly convinced that things haven't really changed. And the memories are there, memories of the good old days when we could all stop at the pub on the way home to dinner.

But Open Space is being crossed, and on the other side there is appearing a marvelous new world, superficially dominated by High Technology, but actually the home of a very different consciousness which is congruent with an open, interactive, global environment, where constant learning is the essence of staying in business. Like it or not, the InterActive Learning Organization has been (or is being) born.

SOME LEARNINGS FROM THE MOMENT OF LEARNING

The experience in the United Kingdom is being replicated in many businesses and organizations around the world. If it has not happened in your shop yet, you may consider yourself fortunate or unfortunate. But rest assured, your time will come.

A significant fact about the moment of learning, and the appearance of the InterActive Learning Organization, is that both arrive without permission. No change agents stood at the gate to usher them in, nor did the board of directors meet to grant approval. It happened. Not without pain, confusion, acrimony, and despair, but it happened. The choice, therefore, was not whether to be an InterActive Learning Organization, but whether to BE at all. Now I call that a "bet the company, bottom line decision."

Put somewhat less dramatically, and maybe a little more accurately, the simple truth is that the external environment suddenly hit a "trip point" — the butterfly landed — which literally coerced the shift, and required a new form of being. Not just any way, mind you, but a rather particular one, congruent with the new shape of the environment. We have called that form the InterActive Organization. The choice offered was Ride the Tiger, or get off and face the consequences.

One might conclude that there is really nothing to do, just wait for the inevitable. There is a certain validity to this conclusion, which we will explore later. In the interim, it will be worthwhile to examine the new form of Spirit in further detail, if only to know what we are confronting. Who knows, we may actually enjoy it.

CHARACTERISTICS OF THE INTERACTIVE LEARNING ORGANIZATION

**High Learning
High Play
Appropriate Structure
Appropriate Control
Genuine Community**

These are not a tight set of specifications to be implemented by executive directive. No memorandum of re-organization will ever do the job, for the path forward follows a rather different route.

One cannot "install" these characteristics and thereby become an InterActive Learning Organization. In a curious way, one must first become an InterActive Organization and thereby manifest the characteristics. There is no linear program, PERT charts won't work, and virtually everything ever learned under the heading of the management of change has zero application. However, what may seem to be an insoluble "chicken and egg" problem has a resolution requiring an unthinkable act, with counter-intuitive results. Let go, and it will happen. More on this later.

The characteristics enumerated are not exhaustive and may not be exact. After all the game has only begun. These are the early returns, constructed from experience to date, extrapolated towards a possible future. Consider it pipe dream, or vision. Your choice.

HIGH LEARNING

The InterActive Organization begins in a moment of High Learning, and it is High Learning that sets the character and tone for subsequent development.

When chaos strikes, transformation begins, and Spirit passes on to a new form. The context is radically reset, perspective alters, paradigms shift. Old realities take on new meaning, and new realities, unseen and unknown, because of the way we looked at things, make their appearance. The moment of breakthrough is disturbing and painful. It is also the ultimate expression of High Learning.

High learning is not unique to the InterActive Organization, for all previous manifestations of Spirit began with such a moment. In the previous situations, however, High Learning was something to be "gotten over" and passed beyond, in order to "get back to normal." For the InterActive Organization, High Learning becomes a way of life.

The continuing presence of High Learning is, in part, merely a natural, and to some extent automatic, response to the environment. In the "good old days," when institutional life spans were measured in centuries, and product lifetimes ran for decades, getting back to normal was a viable option. Our present conditions are rather different, and High Learning is no longer a *sometime* phenomenon, with a *long time* intervening. High Learning has become a continuing phenomenon.

It is, perhaps, a measure of the maturity of the human Spirit that not only is ongoing High Learning demanded by the environment, it is also a possibility. We not only have the challenge, but the means (and the hope) of meeting that challenge.

Somewhere in the wisdom of the cosmos (whatever that might mean), there is a correlation between demand and capacity, with demand creating the Open Space in which capacity can be grown and actualized. If so, Einstein was right, but for the wrong reasons. God does not play dice, there really is an order to things. However, the order exists at a rather different level than Einstein may have imagined. The universe is random, chaotic, and indeterminate, *but it is precisely those conditions that permit, and demand, the growth of Spirit.* Or, as we are now coming to see, chaos is not something to be restrained and feared, but rather respected and embraced, as the fecund ground of new possibility, the mother lode of creativity. Perceiving all of this is a mark of maturity. Consciousness is evolving.

High Learning is not only a natural (automatic) response of the InterActive Organization to the environment, it is also an intentional act, a perfectable skill, an improvable way of life. Here, if ever, content (what is learned) must take a back seat to process (how it's learned). We have already talked about the key elements in some detail. They are: *Embrace Chaos, Force Mistakes to Happen, and Do Intentionally What We Seem to Have Been Doing All Along.*

Embrace Chaos

High Learning, as an intentional way of life, begins with embracing chaos. Every chaotic event is pure gold, creating the Open Space in which innovation manifests. The InterActive Organization actively seeks out those moments and welcomes their appearance with the knowledge that, somewhere in all of that mess, there is a new way of doing business, some new opportunity that may be realized. To avoid chaos is to avoid growth, is to avoid life itself. No longer is this a platitude, it is a business fundamental.

With chaos embraced, the essence of interactivity becomes clear and functional. The chaotic disruptions, emanating from the external environment, are trans-valuated from threat to life (business) into the rich material from which new life (better business) may be created. Only with the intentional, ongoing, interaction with the environment can the full benefit of chaos be realized. With this insight, the organization turns outward, in search of growth, as opposed to inward, in search of stability and the preservation of the system (the way of the ProActive Organization).

Of course there are other benefits to be derived from the environment at large in addition to chaos, but when the worst thing that can happen to you is perceived as

a positive opportunity, risk and failure are "netted out" in advance. All your losses are taken up front. This can be painful, but it also means that everything that follows is pure gain. Interestingly, the only risk is not to risk, and the only failure the unwillingness to learn from it. True security is found only in its absence.

With chaos embraced, it is now possible to fully embrace the total environment. Not just the "nice parts," but all of it. No longer a suspicious stranger fighting the forces of nature, but an active participant, we may understand that house rules (economy) are always derivative from the deep structure of reality (ecology). Economy and ecology are not opposed, indeed they never were, except in our own mis-perception. But now we can get the cart *behind* the horse and consciously align economy with ecology, not because it is right and proper, but simply because it is the only way to go, and completely in line with our perceived self-interest. Healing the planet is healing and growing ourselves. It is good for business.

Force Mistakes to Happen

The InterActive Learning Organization not only embraces chaos, it creates a little of its own. Mistakes are moved from the "cost of doing business" category, over to the asset side of the Corporate Knowledge Balance Sheet. And when business settles into the comfortable repetition of the "same old thing," it is probably time to strike out in some new directions, and force some mistakes to happen, even some very big ones.

The interesting thing about life in the InterActive Learning Organization is that much, if not most, of what happens is not new. Paradoxical as it may seem, we have done it all before, although previously the actions and the results were perceived as somehow being "wrong" and counter-intuitive. Innovation always came from a mistake, and was never according to the plan. We've always known that, but never wanted to admit that was so. After all, if that were true, what would happen to our notions of control, good management, and rational planning? So we played a wonderful game of pretend, clothing the emperor in the fine garments of our imagination. New product breakthroughs were retrospectively engineered into the plan, and those that wouldn't quite fit were conveniently forgotten.

One of the great advantages to life in the InterActive Organization is the increase in truthfulness, and the elimination of all the stress associated with living with a lie, or at the very least, a minor fabrication. If the emperor is naked, say so.

HIGH PLAY

It is said that all work and no play makes Jack a dull boy. The same applies to organizations. However, in the case of the InterActive Learning Organization, play is not the opposite of work, nor is it trivial. It is High Play.

One of the more interesting discoveries of those who study early childhood development is that play is a powerful learning mechanism. When children play doctor, nurse, or fireperson, they are not only exercising one of their major assets, imagination, they are also learning roles, and preparing themselves for adult life. When these roles are organized into endlessly complex and involved games, the process of learning continues, and the object of learning turns to the nature of relationship and structure. How does everything fit together? How does it work? How do you make it work better?

To an adult, observing the process, the roles, structures, and relationships pass by in mind-numbing, kaleidoscopic, rapidity. At one moment, the game is hospital, and the players approach the whole business with awesome seriousness. Then, before that game has barely gotten under way, a new one is invented and takes the stage.

Other games observed among children seem to go on for ever. For days and weeks, the game is airplane, and everybody, even the cat, has a part to play, which is played until each one decides to play a different part. It all seems so random, and without purpose. How could such behavior be truly useful?

The triviality of the enterprise is seemingly proved by the stated reason for its continuance (or alteration). If you ask the children why they are doing what they are doing, or why they suddenly decided to do something different, the answer is shockingly simple. It is fun. Or it is not fun. Fun becomes the funda-

mental criteria. And obviously, if it is fun, it can't be useful.
Perhaps.

If we look at play a little more closely, using the terms
and categories we have been developing, it appears that play
creates instant Open Space, or alternatively, utilizes the Open
Space created by some random event. Every time the game is
changed, all the structures and procedures of the preceding game
are wiped away. Suddenly it is all new, all possibility, and the
participants are freed from the tyranny of all that was. Open
Space is created.

In a similar way, play will also take advantage of Open
Space created by some random event. Suddenly a mother or a
father, for reasons best known to herself or himself, says, "Get in
the car." The destination is unknown, as is the purpose. Gone are

> # In play, we are enabled to do in pretend time what is either impossible, unthinkable, or dangerous to do in real time.

all the things one could do when not in the car. Now what? Well, let's play a game.

The learning possible through play not only concerns the roles, structures, and relationships of life, to say nothing of the technology, it also concerns Open Space: what do you do in it, and with it. Play, in short, is very serious business. It is also fun.

In play, we are enabled to do in pretend time (a time/space continuum of our own making) what is either impossible, unthinkable, or dangerous to do in real time. Then, it sometimes occurs that the line between "pretend" and "real" is removed, and somehow the impossible, unthinkable, and dangerous become common practice. This is called innovation and creativity.

The power of play has not been totally lost on the adult world, as the advent of Gaming testifies. We have business games, war games, economic games. Then in a curious twist of words, we find ourselves playing the game of business, the game of war, the game of economics. The line between "pretend" and "real" is a thin one indeed.

Play is also important, and may be the critical factor, in the appearance of breakthrough science, those moments of superb intellectual adventure when the walls of the possible are shattered. More often than not, such moments occur far from the public view in hidden laboratories or when one is just walking along a river bank. Fortunately, some of those moments have been captured for us in a new genre of literature which we might call the scientific adventure story. Joseph Gleich's book, *Chaos,* is a superb example. Gleich describes the emergence of a whole new science, telling a story replete with blind alleys, happenstantial events and the warts and weird places encountered along the road to real breakthrough. It is a strange and exciting world.

The signal strength of this author is his ability to capture the playfulness of the actors. While the whole business of doing science may be approached with deep seriousness, the actual activity has all the earmarks of High Play. The data of experience are gathered, even as children may gather their building blocks. Then with the material in hand, theoretical structures are erected, lived in, loved, hated, torn apart, and re-arranged. There is a joyful abandon in the shattering of theories, which seems to match, or exceed, the pleasure of their creation. When one theory is gone, the process may begin again with all new possibilities. Much as children erect towering edifices with their blocks, only to send everything plunging into ruin, creating chaos and the Open Space for new creation, so also the scientist in the midst of breakthrough.

At any point in time, the casual observer may see steely grit, flashes of ego and anger, stubbornness or worse. A particular theory will be defended to the death... until the next one comes along. It is the same with children. When one is the doctor, and the patient is dying, it is very tense... until the child decides to be a sailor and the patient gets up to play a different game.

Breakthrough science and High Play go together, and when the play stops, the science is impossible. Without play things get too serious, and one may be deluded into the belief that The Truth is possessed. The end of science is the conviction that there is one right way. Then growth stops, innovation withers, creativity is dead in the water, until fortunately, the universe throws in a dash of chaos, and growth begins again. Not because of the theory, but in spite of it.

None of this is new or news. We have experienced it all before, even though we may not be aware of having done so. Wouldn't it be wonderful if we could now do intentionally what we have been doing unconsciously? The truth of the matter is that we can, and the name of that game is the InterActive Learning Organization.

High Play is characteristic of the InterActive Organization, not only because it feels good (is fun), but also because it is required by the environment in which we now live. With chaos an everyday experience, and the forms and structures of our lives constantly shattered and rearranged, there is infinite opportunity for total despair *or* new creation. High Play is the antidote for despair, and the way to new creation.

APPROPRIATE STRUCTURE AND CONTROL

To this point, it may have seemed that structure and control held a very low place in my estimation, and consequently they would have no place in the InterActive Organization. Nothing could be further from the truth. Structure and control are both characteristic of the emerging organizational form, but the key word is *appropriate*.

In the early days of the movement, which we might now call Organizational Transformation, there was a burning question: What is the structure of the transformed organization? Many opinions were offered, but the conventional wisdom preferred something that was round, flat, decentralized, and networked, marked by high levels of participatory decision making. Hierarchy was definitely out. Asked my opinion, my response was: Anything that works.

I confess to a certain, puckish delight in opposing the conventional wisdom. There was also a method in my madness, for it seemed to me that the issue was never, the *right* structure, as if there were only one, but rather the *appropriate structure*, as measured by its capacity to do the job in a comfortable way. I could even conceive of circumstances where the old bugaboo, hierarchy, combined with autocratic dictatorship, would still be useful. For example, if a fire breaks out in your building, that is probably not the time for a protracted, participatory management discussion regarding the best methods for putting it out. Give me a good old hierarchy with somebody as chief, and let's get on with the job. It's appropriate.

There is plenty of structure in the InterActive Learning Organization, and usually a multiplicity of structures operating simultaneously, with new ones emerging constantly, and old ones passing away, as appropriate. From the point of view of those raised in ProActive Organizations, this will seem somehow "not right," and definitely confusing. But the truth of the matter is that this, too, is a "no news" situation. This is the way it has always been, even though we tried to pretend it wasn't so. In the old days we had the official organization, which was the way things were *supposed* to work. Then we had the "informal organization," which was the way things *did* work. The interesting thing was that nobody really got confused, except those folks charged with drawing lines and boxes, otherwise known as "org charts."

We have always been designing multiple structures to get the various jobs done. Sometimes they were formal, sometimes informal, and the only time we really got in trouble was when we placed greater importance on the structure than on the job and those who were supposed to do the job. Then we had unhappy customers, who found that the product they ordered couldn't be made, not for technical reasons, but because, "It just isn't done that way." Dedicated employees were none too pleased either when they found themselves spending more time running the business than doing the business. That is the sad end of organizations which turn inward and see the preservation of the system as the most important goal.

The ultimate sadness, in this case, is that the natural order of Spirit and structure is reversed. As long as structure provides a useful highway, upon which Spirit may travel, all well and good. But when Spirit is forced to go in unintended directions, it usually goes off the road.

Insisting on the primacy of Structure, even when inappropriate, appears justified on the grounds that structure takes a long time to create, and must be preserved. There is some truth in this, but usually only when we are not clear what we want to do. Then arguments about the "right structure" can, and do, go on for ever, if only to avoid the more necessary, though painful, discussion of what we really want to do. However, *when purpose is clear, structure just happens as a natural expression of that purpose.*

Dramatic proof of this appears in the natural experiment run over the past several years, as we developed Open Space Technology. Leaving the details for the next chapter, the bottom

STRUCTURE HAPPENS!

line result is that large groups of people (current world record is 420 folks) can self-organize a multi-day meeting in less than an hour. In doing so they create a structure that integrates time, place, topics, participants, and leadership for more than 75 workshops or work groups. The resulting structure is more complex than any planning committee could ever imagine, or certainly dare to propose. And it is all done in a fraction of the time that everybody "knows" is required for such work. The prime, and necessary, condition is that everybody knows what they want to do, and are in agreement that it should be done. When purpose is clear, structure happens. If structure doesn't happen, don't keep fussing with it. Go back to purpose.

Appropriate, multiple, changing structures are characteristic of the InterActive Learning Organization. The present environment absolutely demands this approach. Just at the moment we think we have the right people, in the right structure, doing the right job, chaos strikes, and we are forced in a new direction. These conditions are enough to turn the few remaining hairs on the heads of ProActive executives white. Massive states of frustration, stress, to say nothing of despair, can be predicted. However, with a good dose of High Play, all under the heading of High Learning, impending disaster becomes marvelous opportunity. After all, when the old structure goes we can playfully create a whole new game. Life in the organization becomes fun, as it should be.

Appropriate Control

Appropriate structure creates the conditions for appropriate control. And control is as necessary in the InterActive Learning Organization as it was in any of the predecessors. Somebody has to keep the troops on time, and on the road, which means that managers, really good managers, are essential. They must control, but appropriately.

Control and structure obviously go together. Given the latter, you can achieve the former. However, if the structure is inappropriate and therefore arbitrary, control, exercised within that structure, will also be arbitrary. That is where trouble begins. Rarely, if ever, do people object to authority in principle. What is objectionable is authority perceived as unfair, inappropriate, and out of "sync" with the best interests of those involved.

There are some important differences in control practiced in the InterActive Organization as contrasted with prior incarnations. Control becomes a *sometime* thing, assumed and released as structures come and go. Control is not a right, nor is it permanent. It pertains only to a particular structure for a particular time and task. Obviously those who have defined themselves as "being in control" are going to have a very difficult, if not impossible time in the InterActive Organization. For control is no longer the center of meaning, it is merely an occurance that happens.

In the good old days, when structures lasted for what seemed like eternity, one person could be in charge forever. Indeed, the organizational structure could be confused with the organization itself, and control confused with the meaning of life in the organization. It is a gift of chaos to end the confusion. Structures can be shattered, as they are every day, and the organization (that wonderful quantum of Spirit) not only continues, but improves. This same chaos renders us out of control, but life and business continue, and usually deepen, as we explore realities and possibilities that used to be off limits, out of line, beyond control.

The new face of control is very hard on the old ego, both the collective ego of the ProActive Organization, and the individual egos of those involved. It is precisely this harshness

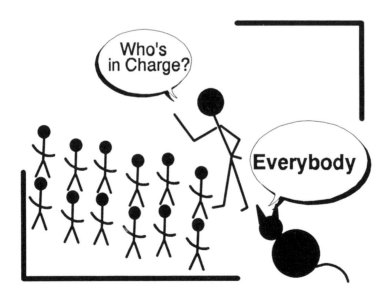

that makes the journey from ProActive to InterActive so difficult to contemplate, for it literally requires moving beyond ego. And if ego is all you have, letting go is no easy task. But that, fortunately or unfortunately, is the name of the new game.

With the absence of ego and permanent control, we are faced with a very interesting question. How do you keep everything together? Where is the center of gravity that can keep the organization whole and functional? Who will be in charge?

The obvious answer: Nobody. But maybe that also means: Everybody.

AUTHENTIC COMMUNITY

The final characteristic of the InterActive Active Learning Organization is real community. Genuine community occurs, not because it is a nice thing (which it is), but because there is literally no other choice. When ego is transcended, because it was kicked out of the way by chaos, and life continues, as it seems to, you are left with what you had to begin, the reality that we are all in this together. Nobody is in charge, we all are. And each one of us bears full responsibility for our brothers and sisters, to the extent of our ability; otherwise the whole ship goes down.

It is not a question of creating community *de novo*, but rather the removal of a delusion that somehow we could exist in lonely isolation as autonomous, closed systems. Chaos teaches the lesson that closed systems do not exist, nor have they ever, except as figments of our imagination.

It is a lesson we are now ready to hear and use. Even though the notion of the *community of all being* has been present as long as humankind has been thinking about such things, it was only a notion, waiting for our maturation before it could be fully appropriated. At the very beginning, we also existed in community, but it was a community of the uninformed and unaware. A homogeneous blob, so to speak.

As the stages of transformation unrolled, the blob differentiated, first in very primal forms, in a ReActive Organization. There were differences that made a difference, but those differences were ones of raw energy reacting to the environment and opportunity. At its best, it was exciting and creative as all true entrepreneurs are exciting and creative. But it lacked a certain finesse, and its rawness often rubbed people the wrong way. We had to learn to be nicer.

Being nicer is nice, but rarely exciting. The Responsive Organization is a nice place to work, and a nice place to do business. That is the up side. It is also terminally dull, lacking critical awareness and intellectual toughness. There is some truth to the phrase, "nice guys finish last."

The gift of the ProActive Organization was to usher in that critical awareness and intellectual toughness. The system was born through the rational interlocking of discrete parts into a well oiled machine. Critical awareness enabled us to isolate the parts, and intellectual toughness brought them together in a functioning unity, whether people liked it or not. As long as those parts were unfeeling cogs and gears, or people willing to be treated as mechanical objects, no problem. But when the folks felt that there had to be a better way, and chaos reared its ugly head to demonstrate that the system wasn't going to work anyhow, something had to give.

Each stage has its own gift. Primal energy, being nice, critical awareness. All significant, but not sufficient. We may now put it all together, and mix well with High Learning, High Play, Appropriate Structure, and Control. That is called Community, the final characteristic of the Interactive Learning Organization. It is the only way to go. More than that, it is fun.

HOW DO YOU GET THERE FROM HERE?

The answer comes in three parts and is shockingly simple. Let go. Let it happen. Make it better.

Easy to say, but somewhat harder to accomplish. Especially letting go. However, if the womb has gone toxic and we are taking the trip down the tubes, there is not much choice. Birth has begun, evolution of consciousness proceeds, and we will take the journey. Of course, we could choose to hang on until the bitter end, and that is exactly what it would be. The bitter end.

Is all of this true? Who knows, but it is my story. Should it also be *our* story, the tale can only get better. One thing, however is clear. We are definitely Riding the Tiger to some-where. While we are waiting for the final destination to be revealed, I propose certain preparations for the journey.

**Let Go
Let It Happen
Make It Better**

Chapter VIII

Let Go, Let It Happen,
Make It Better

It is said that every journey begins with a single step. That step may be the hardest, and perhaps the last, but it is always a single step. It is also intentional and requires "letting go."

Even in those situations where we find ourselves knocked off balance, and stumble towards our future, there exists the possibility of holding on. The result of that action may be disastrous, but the choice remains our own. Hold on, or let go, and get on with the journey.

Robert Tannenbaum, a patron saint for many of us in the field of organization development and transformation, has thought a lot about holding on and letting go. One of his favorite images, which accurately reflects the reality and feelings of the situation, is that of the trapeze artist.

Imagine that you are standing on the small platform high above the ground. In your hand is a trapeze, and far across the open space is another platform, with another person holding a similar trapeze, preparing to swing it towards you. But there is a small hitch. At its farthest reach, the second trapeze will remain just beyond your grasp. In order to get from here to there, it is necessary for you to leave your secure perch, swing out across the void, and then at the farthest extension of your own swing, let go. For a moment you will be suspended in nothingness.

Before you is the approaching trapeze, and below, a painful end of the journey. Your choice.

It is a choice most sane people would rather not make, which I suppose is why there are so few trapeze artists in the world. Yet there are occasions where such a choice must be made.

The odds that we will make that choice are increased considerably if certain conditions are met. First, we must believe that our present perch is untenable, and some of us are coming to that conclusion. Secondly, we must believe that the yonder perch is a better place to be. And thirdly, we must trust the other fellow to swing the second trapeze. Should we anticipate a negative on any of these conditions, it is very unlikely that we will take the trip. Or if taken, it will be done in a half- hearted manner which may be worse than doing nothing at all. It is noteworthy that all we have to go on is belief and trust. No hard data, no guarantees, no certainty. The only way across is by letting go.

Were our present situation as cut and dried as trapezing, the probability that we would quickly get on with the business might increase. But we are not under the circus tent focused on a distant platform. We are asked to risk a whole way of life in return for something we can't quite see yet, and what we can see appears so different from what we've known that we call it "wrong," or at best counter-intuitive. We can think of a thousand very good reasons not to take the journey. Perhaps it is too soon, or not really necessary. Could it be that there is some easier way around the Open Space we face? Some magic carpet we might ride?

Each one of us, and all of us collectively, must make the judgement. In the final analysis, however, the choices reduce to two: Hold on or let go.

LET IT HAPPEN

Should we opt for letting go, the next part should be easy. Like the skier poised at the top of a steep run, when the commitment is made, powers infinitely stronger than ourselves come into play. The shape of the mountain cannot be changed, nor the force of gravity increased or diminished. There is one way to go — down the mountain.

That is the easy part. And it becomes even easier if we align ourselves with those forces, and allow them to take us on our way. This is called surrender. However, should we attempt to "think" our way down the mountain, we find our mental computer overwhelmed by the details of the terrain flashing by. And a decision to stop is almost inevitably an invitation for a spectacular, perhaps disastrous, wipe out.

There is a process in our transformational journey. It is called the cycle of birth and the cycle of death (griefwork). We have been through that cycle before, and presumably we can do it again. The questions are, however, have we learned from our experience, and are we willing to trust the process? Skipping steps is not allowed, and getting off in the middle of the ride inadvisable. Once you let go, it is well to let it happen.

MAKE IT BETTER

The time to prepare for a critical ski run is before you make it. And the time to prepare for the Tiger Ride is before you decide to let go, and take the trip. The situation is complicated by the fact that the Tiger seems to have gotten somewhat of a head start, and so time may be of the essence. But there is still time.

The truth of the matter is that some people have ridden the Tiger before, and more are taking the trip at the moment. We may learn from their experience. Of course, each individual and every organization must take the journey for themselves, and every trip will have its own unique features. But the outline of the journey is well known.

Indeed, the outline of the journey is already part of our experience, if we will but acknowledge the fact. Each one of us has already negotiated the process of moving from infant, to child, to teenager, to adult. Some have done better than others, but we do know the way. While the jump required at the present may appear greater, the process is the same.

To become an adult, one must let go of the teenager. And letting go follows a predictable set of stages. Shock and anger — "They can't really expect ME to get a job?????!!!!!" Denial — "Nothing really has changed, my mother will clean up the mess." Memories — "You remember how it used to be when we were kids?" Eventually we seem to complete the journey, but not without pain and frustration.

As organizations, we have had similar experiences. In the move from ReActive, to Responsive, to ProActive we have known the shock and anger, denial, memories, and the passage across Open Space. Again, some of us have done better than others, and many have refused to make the trip. But the experience is known. We need only to remember the experience and use it positively in our present situation.

Remembering our own experience, and combining that with the experience of others, is part one of "making it better." Remembrance of the experience will not change the shape of the journey nor remove some if its more interesting twists and turns, even as the skier's experience will not modify the shape of the mountain nor the force of gravity. But this remembrance can remove the whole business from the realm of total mystery and terror. We will know, because we have been there before, that each step along the journey has an order: it will come, and it will go. So in the midst of some particularly unpleasant moment, we may be assured that this too shall pass.

We will also know that there is a purpose and reason to the order. The toxic system (womb) sets the conditions for birth. Were it not to occur, we would be condemned to an eternity of being stuck with the way things are. By the same token, the appearance of despair, although exquisitely painful, is the necessary, cleansing cathartic, opening the space for new creation. With this knowledge of purpose, comes the possibility of not only enduring the moment, but actually embracing it, in order to maximize the beneficial result.

Further, we will understand that skipping steps is neither possible nor advisable. To skip over Shock and Anger in the name of social decorum, is only to bury it for later appearance. The shock and anger released at the ending of a significant way of life (manifestation of Spirit) is genuine and must be express-ed, or else it will just come back to haunt us.

Speeding or deepening the process can be done, not by skipping, or shortchanging, steps along the way, but rather by intentionally providing the time, place, and encouragement for the steps to take place. Listening and real presence are required during shock, anger and denial. A more formal ambiance is required for the memories to take place (the Irish wake and its corporate counterpart). We may also ask the question at the critical moment in Open Space: What are you going to do with the rest of your life? In short, there are many things do be done, and roles to play, in order to facilitate the process. For the details, you may refer back to "Notes to Caretakers of Spirit."

Experience can be a wonderful teacher — which leaves a very interesting "chicken and egg" question. There is no problem for people and organizations who have both had the experience (as we all have), and acknowledge it (which some of us find very hard to do). Those folks are in a prime position to use that experience in order to get the most out of the transformational journey, in the least possible time. But what about the others, who may well be in the majority? What about those people for whom the steps of the journey have been treated as an aberrant, unpleasant phenomena, to be endured while getting back to normal as quickly as possible? And once normalcy is achieved, everything is forgotten in the name of not dwelling on problems of the past. How to help them?

Of course, they may be "told," which at least insures that they have heard the words. However, being told something which bears no relationship to your experience (or at least your conscious experience) at best produces amused tolerance, and at worst, positive non-hearing. Knowledge imparted by means of such "telling" is not very useful. So the question remains. How do you provide the experience so that the experience can be provided? Rather like the "first job scenario" in which experience is necessary for employment, but unfortunately employment

155

is necessary for experience, the question is, how do you get from here to there?

The question is important. If the Tiger is moving as we believe, time is of the essence. There are also about six billion of us on the planet. That is an awful lot of people, and not much time. How do you engineer, or facilitate, the essential shift in consciousness?

One answer is to do nothing, just wait and the process will take care of itself. That answer is undoubtedly valid, but not without shortcoming. The cost in human suffering makes it quite unthinkable. In addition, it is very bad for business. Bottom lines tend to get terminally messed up when the Spirit of a place is in constant and continuing turmoil. It would be well if we could get where we are going as quickly, and productively, as possible.

The traditional answer is find yourself an analyst, master, or guru, who will take you through the process. This works, and has worked for millennia. However, I seriously doubt that there are sufficient analysts or gurus, nor the time available in which they may perform their magic. There must be better ways.

Somehow we must all become gurus and analysts for each other, with an approach that is quickly learnable, requires minimal facilities, and support, and yet is effective. It is in this context that I wish to share our experience with Open Space Technology.

I am under no illusion that Open Space Technology is the magic carpet which can take us all to the "other side." Nor is it THE ANSWER, and still less the proprietary product of H. H.

Owen. But it is an approach with a demonstrated capacity to enable large groups of people to experience something very much like an InterActive Learning organization, if not the genuine article. It is also productive, fun, and easy to do.

OPEN SPACE TECHNOLOGY

Open Space Technology (OST) began in frustration, almost as a joke. Such lineage may not commend it for the high purposes we have described. Then again, OST may turn out to be one more example of a breakthrough occurring in ways we would not predict.

The story is this. Ten years ago, I had occasion to organize a rather large international symposium. One year of my life was consumed by endless committee meetings, logistical details, bruised egos (my own and others), and a host of other events too painful or trivial to be remembered. Finally the great day came and went. The effort was declared successful, even wildly successful. BUT. When we looked back on what had taken place, the parts remembered with the greatest pleasure and profit were not the papers and panel discussions, workshops and featured speakers. It was the coffee breaks. The many papers and presentations, elegant as they may have been, could easily have been mailed out in advance. It was the coffee breaks, the open spaces, that provided the significant moments.

It seemed to me that somebody was trying to send a message. If the only thing that really made a difference were the open spaces created by coffee breaks, why bother with all the rest? Why not create one, big open space? The super coffee break.

The idea might sound outrageous, but in fact it was not new. Leaderless and agendaless meetings had been experimented with for a number of years through such things as encounter groups, with the work of the National Training Laboratories (NTL) being a primary example. My experience, however, suggested that although an enormous amount had been learned through such experiments, they seemed to lack a certain practicality. When all was said and done, the question remained: What could you actually do with it all? Obviously one learned a great deal about group dynamics, but somehow the next step towards application never quite happened. What we needed was a super coffee break, which was productive.

The model for OST, so far as I was concerned, originated in West Africa with the Kpelle people, a relatively small tribe living in the interior of Liberia. I had the privilege of living with these folks in the village of Balamah, off and on, over several years, and participated in a number of their major celebrations. What struck me particularly was the powerful, organic nature of these events. It was almost as if the whole village became a single creature, and it "breathed," in a manner of speaking. Even more remarkable was the fact that there seemed to be no organizing committee, or planning as we would understand it.

Nevertheless, everything happened as it was supposed to, and all the players knew, and accomplished their tasks.

The village itself was physically organized in a rough circle, with a central open space. During times of celebration, events started (dancers, drummers, and singers) on the periphery, and then flowed to the center, where they concentrated and grew in intensity. Then, at a certain point, they flowed back to the edges, only to return again. It was breathing in and breathing out. It occurred to me that the village was using the primal elements of the *circle* and *breath* as the essential organizing mechanisms, and context-setting devices. The beauty was that nobody needed to explain the rules, and with the context set, everybody could do what they wanted to, or needed to, and it all worked together. The sense of freedom and responsibility was

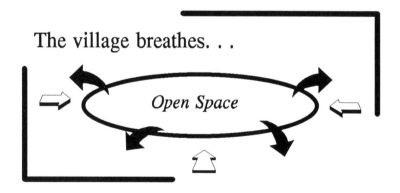

The village breathes. . .

Open Space

incredible, and in no sense did these elements oppose each other. People were freely responsible to do their part. This magic, if magic it was, had been created by the intelligent use of very primal mechanisms. I felt there was another message here. If we could use those primal mechanisms, and/or discover some others, we might well create the productive super coffee break I had in mind.

In creating a meeting, or organizing any other form of productive work, at least three elements are necessary. Identify the task. Create the context. Establish the flow, which means handling time and space in a way that is supportive to the task, comfortable in the context, and agreeable to the participants. In our wisdom, and in the name of efficiency and effectiveness, these elements are typically handled by one individual, or a committee.

Notably absent in this process are the participants and the unique insights, interests, needs, and passions that they might bring. Also absent is sufficient flexibility to meet advantageously any significant event that may have taken place in the external environment subsequent to the original design. Without intending to do so, we often end up creating a design that fits nobody, and deals with nothing (or little) of significance.

This last statement may appear extreme, but I believe it is justified. Our meeting design (job design) is based on averages. The average stomach size (how long between meals). The average bladder size (how long between breaks). The average head size (how much can you pour in at a single sitting). And here is the problem. The average simply does not exist. Each one of us has unique parameters within which we think and work, and our capacity to function effectively is directly related to how well these parameters are dealt with. When it comes to the relevance of the meeting, our recent experience has been that, with the world moving so fast, items of interest today are very likely to be ancient history tomorrow, which means that planning the agenda of a meeting nine months in advance is questionable.

In defense of our present practice, it must be said that consulting all possible participants prior to the design of a meeting is an obvious impossibility. And given the lead time required to organize a meeting of any size, insuring total relevance to the shape of the world nine months hence is also out of the question. It would seem that we do the best we can under the circumstances. However, were it possible to do instant organization, once the participants had arrived, those circumstances would change. The meeting could be designed on the spot, building upon the unique requirements of the participants, and most especially their interests and passions. As we all know, doing what we really want to do is the prime way to insure that what gets done is done well. At least it will hold our attention.

It is here that I thought the insights of my friends from Balamah could be helpful. If we, like they, could find primal mechanisms through which time and space could be organized, congruent with an identified task, nobody would have explain the rules, and the task might be accomplished very quickly. The wisdom of several thousand years had given them two such mechanisms. The circle and breath. The circle defined the space and context, and breath, the flow, timing, pacing.

So far, so good, but we obviously need to fine tune the arrangements, with the designation of more specific times and spaces suitable to the people involved and the task at hand. And all of that will have to be done quickly, or we are back to the planning committee and nine months' lead time.

In the case of the village celebrations, that fine tuning is provided by the ancient rituals which assigned roles and sequences (who does what, when). But there are also two other mechanisms from village life which have the capacity to delineate roles and sequences. The community bulletin board (or the equivalent in a non-literate society, the town crier or talking drum) and the village market place. The former identifies the issues, interests and passions of the people involved, and the later provides the mechanism for getting them all together.

If you have never been to a village market, the shopping mall is the modern equivalent, but in both places, buyers and sellers get together to make a deal. Superficially, the items in trade are goods and services, but there are other items as well. Ideas and relationships, for example. To the casual observer of a traditional village market, the immediate impression is one of chaos, and indeed there is a lot of that. But in spite of the chaos, or I would argue because of it, it all works. I am not at all sure how it works, but I do know that if I am in search of some article, and somebody is selling that article, we will get together.

> # ***Whoever comes is the right people.***
> # ***Whatever happens is the only thing that could have.***
> # ***Whenever it starts is the right time.***
> # ***When it's over, it's over.***

Actually there are some basic principles, and one law. The Four Principles may be summarized as follows: *Whoever comes is the right people. Whatever happens is the only thing that could have. Whenever it starts is the right time. When it is over it is over.* To the logical, western, proactive mind, these principles may seem like nonsense. At the same time, if we are honest, we will probably admit that these principles describe the way life is in the organization — like it or not.

In the village market place, the first principle (Whoever comes...) keeps attention on the fact that you can only sell to the people who are there, and everybody there is a potential customer. In short, there is no need to worry about all the people who are not there, for that is only a distraction.

The second principle (Whatever happens...) keeps you focused on the here and now, for that is the only way you will take advantage of the opportunities presented. Everything that didn't happen, could have happened, or might have happened is beside the point.

The third principle (Whenever it starts...) gets you into the rhythm of things, and reminds you that buyers buy when they are ready to buy. Your wishes, plans, and hopes ultimately make no difference.

The final principle (When it's over...) saves a lot of time and agony. When buyers stop buying, that is it. Time to do something more useful, and don't worry about it.

The One Law is most important, for it keeps clearly in view the essential voluntary nature of the whole enterprise. I call it *The Law of Two Feet*. The point is, everybody has them, and is responsible for their use. In the language of the market place, "There is a time to talk and time to walk." Sophistication is defined as knowing when.

EXPERIMENTAL OPEN SPACE

With the basic mechanisms, the Four Principles, and the One Law all in hand, we began the experiment. The first run

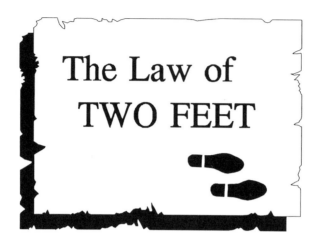

The Law of TWO FEET

occurred in Monterey, California, at a follow-on meeting to my previous effort. This time, however, there was no year-long preparation. In fact, total pre-planning for the meeting (with the exception of negotiating hotel accommodations) took zero time. A theme was announced, potential participants invited (85 actually came), knowing in advance only where the meeting would take place, when it started, when it ended, who was coming, and the nature of the theme. The group successfully designed the gathering in two and one-half hours, and proceeded to self-manage their creation over the succeeding four days. Successive iterations of this gathering, over the next several years, became the test environment in which we developed the approach.

There is no such thing as a standard Open Space Meeting, for each time the approach is used, the details and subject matter vary widely. However, we have found a way to set the scene which works, and over time, certain patterns and predictable results have emerged.

We begin in a circle, for it is important that each person be able to see all the others, and the circle graphically represents the Open Space. The theme is stated, and each participant is invited to identify some area or issue, related to the theme, for which he or she has real passion (not just a "good idea"), and will take personal responsibility. Personal responsibility means convening the group and facilitating the discussion. A title for the issue is written on a large piece of paper, and announced by the participant from the center of the circle. Following that, the participants go to a large blank wall (our Community Bulletin Board) and post the paper. This process continues until nobody has any additional item to post. The point is made: If your item isn't up on the wall for discussion, and you feel left out, there is nobody to blame but yourself.

When all the items have been posted, the Village Market Place opens, and everybody is invited to view the Community Bulletin Board, and sign up for as many issues as desired. From this point forward, the community is totally in charge of its future, and everyone must take personal responsibility for their own learning.

It sounds simple, and it is. In most cases it takes less than half an hour to explain the procedure to a totally new group (slightly longer if multiple languages are involved), and less than one hour to get everything organized to the point that business can begin. The time required for organization shrinks from one year of agony to less than an hour. If you would like to try Open Space Technology in your own organization, please consult the Appendix for further details on procedure and logistics.

EXPERIENCE TO DATE

Open Space Technology has been used successfully in Europe, Latin America, Africa, India, and the United States. Groups have ranged in size from 10 to over 400, and it is not uncommon to have several languages in use simultaneously. An exceptionally broad range of professional backgrounds, educations, ages, cultures etc. can be accomodated at the same meeting. There are circumstances in which OST will not work and should not be used, but the reasons have nothing to do with the rapidity of organization or the lack of planning.

By way of example, 420 teachers, school board members, and administrators organized a day and one-half meeting around the theme, "Education for America," in 45 minutes. Afterwards, they evaluated the quality of their gathering on a scale of 1-10 against "their wildest expectations," and gave it an average score of 9.3.

In South Africa, a group of 75, including the white mayor of Capetown and black leaders, addressed the issue, "Learning

Opportunities in the New South Africa." While most of these people knew of each other, few had actually spoken face to face on issues that really mattered. They self-organized, and completely self-managed (no external facilitators), a day- long meeting, dealing with such "gut issues" as fear, anger, and land reform. At the end of the day, they stood silently in a circle, holding hands, and crying. As one person said later, "We are the New South Africa, and we have a lot of work to do."

One Hundred top executives of the largest Hotel Group in the world (ACCOR) meet in France to build the unity of their corporation. With a broad range of products (otherwise known as hotel chains) from five stars to no stars, operating in some 40 countries, the fact that the group was seemingly united only by a common balance sheet gave pause for thought. Somehow one needed to create a unified core which honored the diversity. Esprit d'corps if you will, but in this case, the theme was "Building Esprit d'ACCOR." Self-organization and self-manage- ment operated as usual, and at the end of two and one-half days, the group walked out with a 150-page document detailing the concrete steps they proposed to take. In fact the steps were already being taken before the meeting had closed because all the decision makers were in attendance. It did not escape the participants, however, that even though the document was impressive, it was, and would remain, only a paper record. The real accomplishment of the gathering was the gathering itself. Somehow Esprit d'ACCOR was no longer a thing to just talk about. It had happened.

One hundred and seventy people representing 28 countries, speaking 17 languages, including an ex-president, and just plain planetary citizens, created a five-day gathering around "Building Global Unity." One hundred and fifty chemists, managers, and support people from a major U.S. Corporation used Open Space to rebuild their organization. Three hundred executives from the U.S. Forest Service did the same thing. The list could be extended considerably, but the basic story would be the same. Given freedom, responsibility, and Open Space, diverse groups of people can do incredible things.

WHAT HAVE WE LEARNED?

If the numbers of folks and practical accomplishments are impressive, all that turns out to be the least noteworthy when contrasted with the quality of spirit manifested in each of the gatherings. Somehow, some way, something truly inspiring takes place with regularity. It begins to look very much like an InterActive Learning Organization.

High Learning

An Open Space event is often experienced as life changing, a "moment" of profound learning. It was never our intent to create such a moment, and I am not at all sure how one would go about doing that. But I do know that moments of High Learning are not uncommon. It doesn't happen for everyone, and the experience is different in all cases, but it does happen.

Typical is the experience of the young woman from India, the managing editor of *Business India*, who reported that, contrary to her expectations of having a vacation and maybe getting a story, she had no vacation, got an incredible story, and "it changed my life." I can put no content on that "life change," never having asked the lady in question exactly what she meant, but others at the same meeting had similar experiences and were more specific. There was a collective realization that the wisdom of India, vaguely and improperly referred to as "religion," had direct and practical application to the business world. In a word, learning in the organization was nothing more or less than the evolution of human consciousness. While westerners might say it differently and perhaps feel uncomfortable with the particular words used, the central realization was that the realm of Spirit and the world of business were not opposed. In fact, business is an expression of Spirit, and becomes more powerful and effective when that expression is realized and built upon. That is a shift, a change in paradigm if you will. It is also High Learning.

The nature and quality of the learning experience is not always so dramatic, but it may be equally profound. A participant from Holland at one of our earlier conferences said that he had never learned so much, so fast, and so well. What captured his attention was that somehow, whenever he was ready, the right combination of people, conversations, and situations conspired to bring him exactly what he was looking for. This gentleman was a European academic of the first order. His previous experience (and therefore his expectation) was that learning takes place in an ordered, linear sequence, which may be violated only at the risk of learning. His experience in Open Space was quite the opposite, which he perceived to be counter-intuitive and wrong. Learning (in this case substantive, practical learning) took place apparently by happenstance, in the midst of chaos, with the most unlikely people (non-experts).

High Learning of a different sort is captured in the comments of an American participant in a summer Open Space gathering sponsored by ASTD (American Society for Training and Development).

> "After presenting or attending hundreds of seminars and conferences, I've become awfully jaded and hard to please...Your camp was a true exception. It exceeded all my expectations. Or I should say, it trashed my expectations and substituted meaningful dialogue, freedom of thought, marvelous new notions, perilous derring-do, and a safe haven cocoon of friendship and warmth in which stretching one's comfort zones becomes easy."

If there is any single quality of an Open Space event that strikes a first-time participant as delightfully strange it is playfulness. Even in undertakings of profound seriousness, (searching for global unity, re-designing the corporation, exploring the new South Africa) there is fun. Nor is it a case of "first work and then fun." It turns out that the work is fun, and the fun is work.

In the Open Space event for the ACCOR group, it happened that the Grand Prix races were being run in the afternoon of the first day. Since the schedule and agenda were both the creation and responsibility of the participants, a number of those participants decided that where they needed to be was before the TV set enjoying the race. One of the more "goal directed" participants found this to be a highly dubious activity, and breathlessly informed me of the supposed truancy. He then asked what I was going to do about it, and I said "Nothing. Everything is working as it should." He left scratching his head, and I am sure thinking some unkind thoughts (Open Space is not for everybody), but I was right.

After the race was over, one of the "watchers," who was also president of one of the hotel chains, told me that wonderful, and unexpected things had happened. His group had gone to the race because they were totally frustrated by the results of their discussions. They had been getting nowhere. But somehow in the middle of the race, the mental logjam broke. It was not only a question of taking a break and coming back with a fresh perspective, the competitive environment of the race had actually provided the inspiration for the new ideas that literally poured forth. Moreover, his folks continued to play "race car driver." as they worked out the new schemes. I never achieved total clarity about the nature of that particular game, due in part to my limited French. But there was no question about the presence of High Play, and the fact that it had contributed enormously to the solution of some rather complex issues.

Appropriate Structure and Control

When people first hear about Open Space Technology, it is often believed to be without structure and control. It is quite true that the process begins with a blank wall, and therefore with no formal structure or agenda; without those, it is perhaps a little difficult to see how control might be exercised. This is a classic case of the deception of appearances.

While there is no original formal structure, there are boundaries, for Open Space without boundary is an impossibility, logically and practically. Boundaries are created by the theme, start and stop times, the Four Principles, One Law, Community Bulletin Board and Village Market Place. By design, these boundaries are quite loose, intended less to tell people what to do, than to create the environment in which they may figure out what needs to get done, with a minimum of hassle. But they are boundaries.

173

With regard to formal structure and control, that exists in abundance, usually commencing forty-five minutes after the event gets under way. By any reasonable estimate it is a more complex and inter-related structure than any planning committee would design or propose. In large Open Space event, involving hundreds of people, there may be as many as 75 workshops organized over a four-day period, all neatly arranged with conveners (leaders), locations and times. Unlike the structure of a "standard" meeting, however, all of the above emerges from the people as an expression of their interests and passions. In short, it is appropriate and sensitive to the needs, desires, capacities, and cultures of the people participating.

Cultural sensitivity is a subtle strong point of Open Space Technology. Typically, when multiple cultures convene for some reason, there are endless conflicts around time and space. North Americans think South Americans are slow and always late. And the South Americans take their North American brothers and sisters as rude, pushy, and too much in a hurry. In truth nobody is late or early, it is just different senses of time, and the uses of space.

When one group is forced to operate in the time/space of the other, real problems emerge having nothing to do with the business at hand. In Open Space, those problems just do not occur. When each person, or group, has both the freedom and responsibility for organizing their own time and space, they will feel comfortable and productive, and if not, they have the automatic license to change things. Furthermore, when it is necessary for different cultural groups to work together, it is always a matter of negotiation, finding a mutually agreeable time and space. But this negotiation is undertaken because the groups in question WANT to get together around a common issue and concern, as opposed to being forced into some arbitrarily imposed format. Of course, such joint undertakings don't always happen, nor are they always successful, but the cause of failure is not a question of time and space. It is quite simply that the level of mutual want was not sufficiently high. Cultural diversity is acknowledged and honored.

Quite the opposite of a lack of structure, there is an abundance, indeed multiplicity, of structures, which are constantly being changed and adapted to the needs of the group. And the group is in charge.

Under these circumstances, control is also exercised by "managers" — those people who have taken responsibility for insuring that a particular work group will happen. But the relationship between the "manager" and the members of the group is always consensual, which is the result of the Law of Two Feet. Anybody who doesn't like it can leave.

The conventional wisdom would predict total, non-productive disorder. As a matter of fact, what takes place is plenty of fertile chaos and the creation of new order. Driven by passion, and in line with the Spirit of the place, structure and control assume a positive role. When 100 senior ACCOR executives from around the world, each used to running their own shop, create and self-manage a two-and-a-half-day meeting, producing

150 pages of proceedings available at the time of leaving, something beyond non-productive disorder has taken place. Such results are not the exception.

Genuine Community

A meeting run in Open Space typically becomes a tight community, marked by High Learning, High Play, and appropriate structure and control. Why this happens, I do not know in precise detail. But it is significant to me that serious "business" gatherings regularly display the kind of community sought, and often not achieved, in workshops organized specifically for the purpose of building community. And all of that takes place without a single exercise or training program. Nor is there a cadre of expert facilitators standing in the wings, or actively managing the process. It just happens, all the more surprising because it was unexpected.

Quite commonly, participants who have attended more than one event, often on widely different subjects, talk about becoming an "Open Space junkie" and "coming home." Doubtless there is some degree of hyperbole here, and the reaction is not universal. But it is clear that having once experienced the creative "highs" possible in Open Space, going to "just another meeting" is not appetizing. As for coming home, I share the feeling. Having now enabled a number of Open Space gatherings, all over the world, with an enormous range of cultures, colors and concerns, I can say quite honestly, I always feel at home. Not at home in a closed, bar the door, keep out the world sense, but at home in an openness that welcomes difference and celebrates diversity. It is a place where strangers become non-strangers remarkably quickly.

At one of our early gatherings, our meeting room was located just down the hall from a free real estate seminar. Over several days, the real estate folks passed our door and looked in, at first with curiosity and then with something approaching longing. I guess they could see that we were having fun and they were envious. By the second day we had drop-ins, and on the final day, I noticed several of the free real estaters in our closing circle.

The closing circle is not a prescribed ritual, by any matter of means, but it seems to result as a natural expression of the community we become. It is not a time for report out in a formal sense, but rather an opportunity to share what the whole event may have meant. I was not a little curious to hear what our drop-ins might say.

When the moment arrived for the first drop-in to speak, he was almost speechless. Regaining self-control, he thanked us for allowing him, a stranger, to join the gathering. It had been, he said, the most significant moment of his life, something he had been searching for. He felt he had come home.

Was that an unguarded, emotional outburst from some unbalanced soul? Possibly. But it is also a common experience in Open Space.

WHEN OPEN SPACE DOESN'T WORK

There are times when Open Space Technology shouldn't be used, and doesn't work. The major contraindication for Open Space is when you have some specific procedure that you want to introduce, and you already 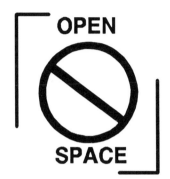 know all the details, and how it should be done. The issue is not to excite the creative potential of the folks involved, it is to equip them with the detailed information and special skills necessary to do the job. This is not the time for Open Space, and should you use it, all you will get is a mess, and not a very creative one.

There are other times when Open Space will not work, having nothing to do with the task at hand, but rather the people involved and their attitudes. Open Space will not work when people are forced to join it. They may be encouraged, cajoled, but never forced. People forced into Open Space will become fearful, negative, disruptive, and unpleasant.

The final circumstance under which Open Space should never be used is when there really is some covert, specific agenda, and those who hold that agenda wish to maintain control. I say specific agenda, because of course every Open Space meeting has a goal or purpose. But it must be general in nature, like "Building the Esprit d'ACCOR," with absolutely no attachment to specific outcomes. The details must remain with the folks, and to attempt control of their progress is to limit their creativity. It will frustrate them and typically make the situation worse, maybe much worse.

Lastly, there is a special group of people who should never attempt Open Space. These are people who define their lives as "being in control." It will not surprise you to learn that I feel they are suffering from some delusion, but that does not alter the case. In Open Space, they will be miserable, and make everybody else miserable.

One might reason from this last caveat that there are certain organizations so control-oriented that Open Space should not be contemplated. In my experience, this is not true. There is a natural self-selection process which screens out the problem cases. Once an organization has considered the issue and voluntarily made the choice, the chances are Open Space will work. At least it always has.

One final note. It is interesting that every single organization with which I have worked, has expressed the opinion, prior to the event, that Open Space may be a wonderful thing — but it will never work in *this* organization. One hour after the event begins, they forget this opinion. All of which makes the point: being nervous about Open Space is quite natural.

WHAT DO WE DO WITH OPEN SPACE?

The uses of Open Space Technology are various. At the simplest level, we can certainly profit from the reduced time necessary for planning meetings, and the increased level of creativity and productivity, both of which translate to direct bottom line results, either as costs saved, or return on investment.

There is a broader usage, which so far as I know, has not been tried. That is to run a whole business on the basis of Open Space. In other words, do what we do in meetings — every day. That may seem a little farfetched, but when you think about it, a lot of business, indeed all organizational life, is nothing more than a succession of meetings, and if all of that could be characterized by High Learning, High Play, Appropriate Structure and Control, to say nothing of Genuine Community, it would be fun to get up in the morning to go to work. Or would it be play?

Actually, this idea is not as farfetched as it might seem. It is my belief that this is where we are headed, and indeed must go, if we are to flourish on this planet. The InterActive Learning Organization, by this or any other name, is positively required by the emerging conditions of the environment. The only issues from my point of view are, how soon can we get there, and how do we do it?

If there is any validity to my story, the way forward is deceptively simply: Let Go, Let It Happen, and Make It Better.

But then we come back to the chicken and egg. How do we experience our future so that we might trust sufficiently in order to let go and experience the future?

Open Space Technology may provide a way. Not the only one, maybe not the best, but nevertheless a way. If Open Space Technology creates an environment which is analogous to, or equivalent with, the InterActive Learning Organization, we have the beginning of our answer. We will have experienced the future, which makes letting go of the past so much easier.

Appendix

A USER'S GUIDE TO
OPEN SPACE TECHNOLOGY

(For a complete introduction to Open Space
please consult our new book, <u>Open Space Technology:
A User's Guide</u>)

THE REQUIREMENTS OF OPEN SPACE

Open Space Technology requires very few advance elements.
There must be a clear and compelling theme, an interested and
committed group, time and a place, and a leader. Detailed
advance agendas, plans, and materials are not only un-needed,
they are usually counterproductive. This brief User's Guide has
proven effective in getting most new leaders and groups off and
running. While there are many additional things that can be
learned about operating in Open Space, this will get you started.
Some material has been included here which also appears in the
book in order to present a relatively complete picture.

THE THEME — Creation of a powerful theme statement is
critical, for it will be the central mechanism for focusing discus-

sion and inspiring participation. The theme statement, however, cannot be a lengthy, dry, recitation of goals and objectives. It must have the capacity to inspire participation by being specific enough to indicate the direction, while possessing sufficient openness to allow for the imagination of the group to take over.

There is no pat formulation for doing this, for what inspires one group will totally turn off another. One way of thinking about the theme statement is as the opening paragraph of a truly exciting story. The reader should have enough detail to know where the tale is headed and what some of the possible adventures are likely to be. But "telling all" in the beginning will make it quite unlikely that the reader will proceed. After all, who would read a story they already know?

THE GROUP — The group must be interested and committed. Failing that, Open Space Technology will not work. The key ingredients for deep creative learning are real freedom and real responsibility. Freedom allows for exploration and experimentation, while responsibility insures that both will be pursued with rigor. Interest and commitment are the prerequisites for the responsible use of freedom. There is no way that we know of to force people to be interested and committed. That must be a precondition.

One way of insuring both commitment and interest is to make participation in the Open Space event completely voluntary. The people who come should be there because they want to be there. It is also imperative that all participants know what they are getting into before they arrive. Obviously they can't know the details of discussions that have yet to take place. But they can and should be made aware of the general outlines. Open Space is not for everybody, and involuntary, non-informed participation is not only a contradiction in terms, it can become very destructive.

This raises the obvious question of what to do with those people whom you want to involve, but who, for whatever reason, do not share your desire. There are two possibilities. The first is to schedule two sessions, and trust that the first one will be so rewarding that positive word of mouth testimony will draw in the recalcitrant. The alternative is to respect the wishes of those involved. In the final analysis it remains true that genuine learning only takes place on the basis of interest and commitment, and there is absolutely no way to force any of that.

The size of the group is not absolutely critical. However, there does seem to be a lower limit of about 20. Less than 20 participants, and you tend to lose the necessary diversity which brings genuine interchange. At the upward end of the scale, groups of 400 work very well, and there is no reason to believe that number could not be increased.

SPACE — The space required is critical, but need not be elaborate or elegant. Comfort is more important. You will need a room large enough to hold the entire group, with space to spare in which the participants may easily move about. Tables or desks are not only unnecessary, but will probably get in the way. Movable chairs, on the other hand, are essential.

The initial setup is a circle with a large, blank wall somewhere in the room. The wall must be free from windows, doors, drapes, and with a surface that permits taping paper with masking tape. The wall should also be long enough so that the total group may stand before it, and never be more than three to four deep. The center of the circle is empty, for after all we are talking about Open Space.

If the room is very large, additional break-out areas may not be required, but they are always helpful. Best of all is the sort of environment in which there is an abundance of common space. If you are going to use a conference center or hotel, find one with plenty of conversation nooks, lobbies, and open grounds, where people may meet and work undisturbed, and without disturbing others.

TIME — The time required depends on the specificity of result you require. Even a large group can achieve high levels of interaction combined with a real sense of having explored the issues in a matter of eight hours. However, if you want to go deeper than that, reaching firm conclusions and recommendations (as would be the case for strategic planning or product design), the time required may stretch to two or three days.

More important than the length of time is the *integrity* of the time. Open Space Technology will not work if it is interrupted. This means that "drop-ins" should be discouraged. Those who come must be there at the beginning, and stay for the duration if at all possible. By the same token, once the process begins, it cannot be interrupted by other events or presentations. These might come before or afterwards, but never in the middle.

THE BASIC STRUCTURE

Although it is true that an Open Space event has no pre-determined agenda, it must have an overall structure or framework. This framework is not intended to tell people what to do and when. Rather, it creates a supportive environment in which the participants can solve those issues for themselves. Minimal elements of this framework include: Opening, Agenda Setting, Open Space, and Conclusion. These elements will suffice for events lasting up to a day. Longer events will require the addition of Morning Announcements, Evening News, and probably a Celebration.

A standard Open Space Design, using all these elements appears below. If the event you anticipate lasts longer than the time indicated, simply replicate the middle day. If shorter, you will find that an Opening, Open Space, and Conclusion will suffice. Generally speaking, the minimum time required is five hours, but that is cutting it rather close.

OPENING — We have found that a very informal opening works well, especially if the group involved is an intact work

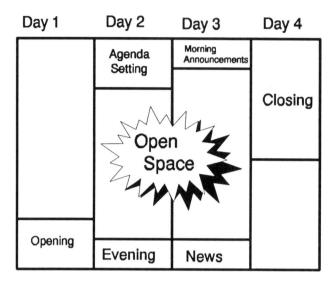

group. An evening meal and a time for catch-up conversation will effectively set the stage. Should the group not have any prior association, the simple device of having all the participants introduce themselves by giving their names and telling a short story from their lives to illustrate who they are will usually do the job. Detailed and involved "icebreaking" exercises do not seem to work very well, and more to the point, set the wrong tone. After all, we want Open Space.

AGENDA SETTING — This is the time for the group to figure out what it wants to do. The details for this procedure are given below.

187

OPEN SPACE — is exactly what the words imply, open space and time for the group to do its business. There is literally nothing here at the start.

ANNOUNCEMENTS — A short period every morning for the group to catch up on what it is doing, where, when, and how. Nothing elaborate, no speeches, just the facts, nothing but the facts.

EVENING NEWS — This is usually a time for reflection and occasionally fun. Not to be confused with a formal report-out session, the approach is "What's the story?" — with participants voluntarily providing the tale.

NOTE

All Events in Open Space are voluntary, however participants should be strongly urged to attend Morning Announcements and Evening News. Gathering the whole community twice daily is essential.

CELEBRATION — If your Open Space event is like all the ones we have seen, particularly multi-day affairs, by the last night it will time to celebrate, otherwise known as having a party. Even in "serious" undertakings like preparation of the corporate strategic plan, when it is over, it is over, and people will enjoy celebrating that fact. We suggest doing the celebration in the spirit and manner of the rest of the event. All of which means don't plan it in advance. It may be worthwhile to have some taped music if your people are inclined to dance, but other than that you will undoubtedly find that the talent you need is already available in the folks you have. Use it. Skits, songs, humorous reviews of what has happened, will amply fill the evening, and add to the learning experience.

CLOSING — We try to keep the closing simple and serious. Simple in that there are no formal presentations and speeches. But serious, for this is the time for announcing commitments, next steps, and observations about what the event has meant. The closing event is best conducted in a circle with no "head table." Start anywhere, and go around the circle allowing each participant, *who wants to*, the opportunity to say what was of significance and what they propose to do. But do make it clear that nobody has to say anything. In very large groups, hearing from everybody is obviously impossible, but two or three folks may be asked to volunteer.

FORMAL REPORTS — The formal report-out session has apparently become a fixture of conference life. However, we find it to be boring and generally non-productive. There is never enough time for each group to say all they wanted to, and if sufficient time is allocated, the majority of conference participants are uninterested at any given time. As an alternative, we recommend using a simple word processing system, a computer conferencing system, or both.

In a recent conference 200 participants created 65 task force reports (a total of 200 pages) which were available as the participants left the conference. Mechanically, all that is required is a bank of computers (low-powered laptops will do) and a request to each group organizer to enter the results of their deliberations into the system. They can either type it in them-selves, or for the "non-typables," a small group of secretaries will do the job. We print out each report as it is entered and hang it on the wall, providing an ongoing, real-time record of the discussions. The obvious advantage here is that participants find out what is happening, as it is happening, rather than waiting until the end when it is too late. Of course, having the proceedings at the end of conference, rather than six months later, is a pleasant and positive surprise.

MEALS — You will notice that meals are not listed on the agenda, nor are there any coffee breaks. The reason is quite simple: once the conference starts to operate in small groups, there is usually never a time when something of substance is not going on. And in accord with the Third Principle, it will take place in its own time. All of this creates a small, but not insoluble, problem for such things as meals and coffee-breaks. Our solution has been to have coffee and other refreshments available in the main meeting room, so people partake when they are ready. No need for the whole group to get into lockstep, and stop an important discussion just because it is coffee-break time. Likewise with meals. We suggest buffets, open and available over a several hour period, so people can eat when they want to. There are two exceptions to the flexible meal/coffee-break schedule: an opening dinner if there is one, and dinner on the last night.

The whole point is that the pacing and timing of the conference must be determined by the needs of the group and its learning process, and not by the requirements of the kitchen.

LEADERSHIP

The leadership of an Open Space event is at once absurdly simple and very tricky. The simplicity derives from the fact that the group itself will, and must, generate its own leadership. The tricky part comes in *letting* that happen. The demands placed upon the initial group leader are therefore limited and critical. Dealing with the limited aspects of group leadership is easiest and may therefore be done first. The functions here are to set time, place, and theme. Time and place are simply a question of where and when, both of which have been discussed above. Setting the theme involves creating the written theme statement describing where the group is starting, and where it hopes to go in general terms.

Now we come to the tricky part. Leadership in Open Space requires a style that some may find uncomfortable and counter-intuitive. This is especially true for those who equate leadership with control. There is no question that when we know exactly what we are doing, and where we want to go (as is presumably the case, for example, in a manufacturing process), tight controls are essential. In fact, control is the very heart of good management. We get into trouble, however, when we understand leadership simply as advanced management, and therefore, if the manager controls, the leader must control absolutely. Sensitive leaders today, in a world marked by progressively expanding Open Space, know all too well that most of what they have to deal with is beyond their control, and maybe out of control.

Leadership defined as control can only fail. But that is not the only definition. Gandhi described the leader as one who intuits which way the parade is moving, and then races to reach the head of it. The function of leadership is to provide a focal point for direction, and not to mandate and control a minute-by-minute plan of action. The details must be left to the troops, which means amongst other things, the troops must be trusted. In no case can any leader possibly solve all problems or direct all actions. Leadership in Open Space requires that one set the direction, define and honor the space, and let go.

There are Four Principles and One Law which serve as guides to the leader and all participants. The principles are: Whoever comes is the right people. Whatever happens is the only thing that could have. Whenever it starts is the right time. When it is over, it is over.

The first principle reminds everyone of the obvious fact that those present are the only ones there. Whatever gets done will get done with them, or not at all. There is little point, therefore in worrying about all those who should have come, might have come, but didn't come. It is essential to concentrate on those who are there. The experience is that, in some strange way, the group present is always the right group.

In more practical terms, it has been discovered that if the group is deeply involved in the issue at hand and excited by the possibilities, that involvement and excitement are contagious, and others will soon join in. Even if the technical expertise present is not of the highest order, a committed group will find the needed expertise. However, if all the time is spent in telling each other that the group is neither right nor competent, it is always the case that the group will live down to it expectations.

None of this is to suggest that effort should not be made before the gathering to be sure that invitations are extended to critical people. Or indeed that those critical people should not be specially urged to attend. However, when the gathering starts, it is unarguably true: those who came are the ones who came. Whatever gets done will be done by them, or not at all. They will be the right people.

The second principle is yet another statement of the obvious. Given the theme (job) at hand and the people in attendance, whatever happens is the only thing that could have. Change the people, time, place, or theme, and something different will result. It is, of course possible that the result of the gathering could be a miserable failure, but experience shows that such a negative result is usually the product of negative expectations. Expect the worst, and you will very often get it.

Expectations are in fact critical. Be prepared to be surprised — positively. Those who come to an Open Space event with a precise and detailed list of intended outcomes will be frustrated. More than that, they will inevitably miss the positive and useful things that occur. Never before, and never again will the assembled group gather in that time and place. No one could possibly predict the synergism of effect that will take place when those particular people assemble. Some of what happens will be non-useful. But it is the special function of the leader to raise the expectations of the group, and heighten their sensitivity to the opportunities at hand, whatever they may be.

Here is the most difficult and important point about leadership in Open Space. *The leader must truly trust the group to find its own way.* Attempts on the part of the leader to impose specific outcomes or agenda will totally abort the process. Any person who is not fully prepared to let go of their own detailed agenda should not lead.

The third principle will seem essentially wrong to those whose lives have been dictated by the clock, which is basically all of us. The conventional wisdom says that if you want to get something done, you must start on time. The conventional wisdom is right so long as you know what you are going to do, and how. On the other hand, when creativity, and real learning are involved, the clock can be more of a detriment than an assist. Things will start when they are ready, and whenever they start is the right time. In fact, when the creative learning moment arrives, it seems to create its own time, or put another way, clocks don't seem to matter much anymore. The Open Space environment provides the nutrient setting for creative activity, and those who would lead in that environment must keep their eye on the creative process and forget about the clock. When "it" happens, it will happen in its own time, and scheduling a break-through for 10 am is not only an exercise in futility, it is consummately destructive of Open Space.

Open Space Events do, of course, occur in time, which means that there must be a time of beginning and a time for closure. But everything in the middle must be allowed to run its own course.

The final principle, "When it is over, it is over," again states the obvious, but it is a point we may forget. Deep learning and creativity both have their own internal life cycle. They may take more or less time, but when they come to completion, they are over. Occasionally this means that we have to spend more time than we had planned, but more often than not, the reverse is true. The creative moment has a nasty habit of occurring very quickly, and just because the session or meeting was scheduled to take two hours is no reason to sit around and waste time after the moment has passed. When it is over, it is over.

Finally we come to the One Law of Open Space. It is a law only in the sense that all participants must observe it or the process will not work. We call it the Law of Two Feet. Briefly stated, this law says that every individual has two feet, and must be prepared to use them. Responsibility for a successful outcome in any Open Space Event resides with exactly one person — each participant. Individuals can make a difference and must make a difference. If that is not true in a given situation, they, and they alone, must take responsibility to use their two feet, and move to a new place where they can make a difference. This departure need not be made in anger or hostility, but only after honoring the people involved and the space they occupy. By word or gesture, indicate that you have nothing further to contribute, wish them well, and go and do something useful.

WHEN NOT TO USE
OPEN SPACE TECHNOLOGY

As there are individuals who should not lead in Open Space, there are also situations in which Open Space Technology is not appropriate, and in fact may be counter-productive. Open Space Technology is effective when real learning, innovation, and departure from the norm are required. When you aren't quite sure where you are, and less than clear about where you are headed, and require the best thinking and support from all those who wish to be involved, Open Space Technology will provide the means.

On the other hand, if the present state, and future position are crystal clear, along with all the intervening steps, Open Space Technology is not only a waste of time, it will be very frustrating. Using a very mundane example, if the task at hand is the implementation of a known technology, such as a word processing program, or an established office procedure, inviting people to be creative and inventive is quite beside the point. They simply have to learn the skills and methods required. There is no mystery. Just do it.

CREATING OPEN SPACE — Introducing The Event

With the preamble out of the way, it is time to get on with the event. What follows is a walkthough of the format we have used. But please do not treat it as an unchangeable script. The needs of your group and your own style will ultimately determine the best way.

If this is a "first time" for you as a leader, we strongly suggest that you take a practice run through. Start by becoming completely familiar with the walkthough material provided below. Imagine that you are actually leading a group, and read through the script. Do this until you don't need the script, and then go one step further. Forget the script's words and use your own. At that point, you are probably ready for a real group, but don't make your first effort "the critical one." Find some friends and colleagues who are willing guinea pigs. They should have fun, and so should you. As a matter of fact, having fun is the key indicator that you are ready to take on a group for real. If it isn't fun, don't do it. Maybe you should never do it, or maybe you just need more practice. But HAVE FUN.

Assume that your group is now assembled in a circle, with a large blank wall behind them. You walk into the center and begin:

"Our theme for this gathering is _____. In the next __days, we are going to develop our best thoughts around the issues and opportunities associated with our theme.

As we start, I want you to notice the blank wall. That is our agenda. Just out of curiosity, how many times have you ever been to a meeting where the agenda was a completely blank wall?

If you are wondering how you ever got into all this, or even more, how you will ever get out, you should know that while Open Space Technology is a new approach, it is not untried. Groups all over the world, some as large as 400, regularly create their own agendas for multi-day meetings in less than one hour. They then proceed to self-manage the whole affair. While this is not a contest, there is no reason for you to do less well than those who have gone before you."

It is worthwhile to pause a moment here. Let them look at the blank wall and really understand that there is no agenda except as they make it. Some people will beginning to feel rather nervous, and others will be demonstrably so, but nervousness (anxiety) at this point is a plus, for it represents available energy or spirit just waiting to happen. The art is to wait long enough for it to build, but not so long that people will question what they are doing, or worse yet begin a discussion about the whole process. If that sort of discussion begins, you will have lost the moment. So pause for a moment, and then move on.

"To get 'from here to there' we will use two very simple mechanisms — the Community Bulletin Board, and the Village Market Place. In a few moments, I will ask you to identify any issue or opportunity you see around our theme, give it a short title and write that down on the paper provided. Then stand up in front of the group, say what your issue is, and post the paper on the wall. Make sure that you have some real passion for this issue, and that it is not just a good idea for somebody else to do. For you will be expected to take personal responsibility for the discussion. That means saying where and when the group will meet, convening the group, and entering the results of your discussion into our computer system (if you are using a system). You may offer as many issues as you like, and if at the end of the day, you do not see your issue on the wall, there is exactly one person to complain to. Yourself.

199

"Once all the issues are up, we will then open the Village Market Place, and everybody will be invited to come to the wall and sign up for as few or as many of the groups as they desire. From there on out, you are in charge.

"Even though Open Space is truly open, there are some principles and one law that we need to keep in mind. The Four Principles are. . . [see above]. And the Law is what we call the Law of Two Feet. Everybody has them. . . . [see above]."

We find it helpful to write the Four Principles and One Law out on a large piece of paper which may be hung on the wall for future reference.

"Keeping the Four Principles in mind, along with the One Law, it is now time to get to work. Along that line, there is one question to start. *What are the issues and opportunities around our theme, for which you have real passion and will take genuine responsibility?*

"And when you have identified an issue or area, give it a short title, write it down on the paper provided and sign it. Leave some room at the bottom for others to sign."

If you have a relatively large group (25 and up) it is helpful to have paper available in a pile in the center of the circle. A basket of magic markers will also help. The paper should be large enough so that when taped to the wall it may be easily read by the group from a distance of about 10 feet.

"As soon as you are ready, stand up where you are, read out your title, and tape it to the wall. Don't wait to be asked. Go when you are ready."

Keep on going until everybody with a subject they want to work on has posted it on the wall. There will be a certain amount of noise and confusion, which is positive and good, but keep it down a little bit so people can hear. Most important, don't let the people start to discuss any of the items at this point. There will be plenty of time for that.

When it seems that all the items have been posted on the wall, ask if there are any more, and direct the group's attention to the wall. If your group is like all others we have worked with, the wall should now be covered with things to do. You might say something like:

"For those of you who wondered whether we would have something do, you might take a look at our wall. You might also note that we have generated the items for our agenda in less than _____ minutes.

"Our next act is to figure out who is going to do what, when, and where. To move that business, I would ask that every person who has an item on the wall go up and write down the time and place where your group will meet.

"Make sure your name is on the paper. For example, your group will meet from 10 to 12 o'clock in conference room C, or maybe out by the swimming pool. Space is on a first come, first served basis."

Prior to this part of the program, the leader should post a list of available meeting places. Of course if you are the only group meeting in a conference center, the space problem is simplified, and the groups can meet anywhere they feel comfortable.

"Don't worry about conflicts. We'll work all that out in a moment. Once you have selected a time, move your paper to the appropriate part of the wall. If you want to meet early, put it on the left side. For the end of the meeting, put it on the right. Those who want to gather in the middle, put it in the middle."

For meetings lasting longer than one day, it is helpful to divide the wall with tape into as many sections as there are days. You might also post *Morning Announcements* with a time at the beginning of each day, and *Evening News* just before supper. Additional time demarcations are not needed, and tend to get in the way.

"While they are doing that, all the rest of you might also stand up and take a look at the various offerings. When you find one that interests you, where you could learn or contribute, sign

your name on the bottom. Sign up for as many as you like, and don't worry about conflicts. We'll take care of them in a minute. It may seem a little chaotic, but it turns out that chaos is the way the fields of the mind are plowed so that new ideas can grow."

At this point, things are likely to get pretty noisy, and some might say chaotic. Leaders used to having things happen in relative silence, and in strict order may get very nervous. It is all right to be nervous, but don't try to straighten things out. THE GROUP WILL TAKE CARE OF ITSELF. A little chaos at this point is a good and necessary thing. First of all, everybody probably needs a stretch and some conversation. But most of all the rising noise level is a positive indication that the group is getting to work, and good things are happening.

Let the group bubble along for a few minutes, but before long some people will start to experience conflicts. They want to go to two different groups which are supposed to meet at the same time. Or two different groups are scheduled to meet in the same place at the same time. When you sense this happening, or even if you don't, stand up and get the group's attention. You may have to raise your voice, but noise won't hurt.

"Some of you may be finding a few conflicts, but it should be easy to work out. It is called negotiation. If you want to go to two groups meeting at the same time, find the group leaders and see if you can get them to merge their sessions or

change their times. Of course, if they won't do that, you will just have to make a choice, but that is the way life is.

"From here on out — you are on your own. As soon as your group is ready to go to work, go to it. We'll see you all back here at _____ [Evening News, Closing Session, whatever]."

We find it very useful at this point for the leader to actually leave the room, if only for a cup of coffee. It really makes the point, as little else could, that each individual and the total group are now responsible for what happens.

From here on out, the role of the leader will be infinitely less visible, but very important. There are, of course, several specific things to be done, such as convening the Morning Announcements, Evening News, and the Closing Session. The leader should also make arrangement for the collection of the conference out-put if a record of formal decisions and deliberations is required. This may be as simple as having flip-charts transcribed, or even better, insuring that the computer conferencing system is up, accessible, and used.

One most important function on the leader's to do list is to take care of the room and the wall where the created agenda lies. We find ourselves spending a lot of time in that room, sometimes doing little more than picking up coffee cups, or re-taping agenda items that have come unglued. This may seem trivial and non-useful, but at the symbolic level it is a powerful statement of the leader's concern for the common space. On a more practical level, it usually turns out that the initial meeting room plays the role of "Mission Control." It is the place where everybody, sooner or latter, drops by to see what's happening or where to go next. Simply by being there, it is possible to keep tabs on how everything is coming along.

The major function of the leader, however, is not to do certain, specific things, but rather to sustain the atmosphere of Open Space. In most cases this involves little more than walking about and seeing how things are going. When difficulty is encountered, it is important not to take charge, but rather to throw responsibility back on those who need to hold it. For example, it is not unlikely that one or two of the participants will find themselves slightly lost, and come to the leader expecting to be told what to do. An appropriate response would be a question — "What would you like to do?" followed up with the assurance that there is nothing wrong with doing nothing. It may just be that some individuals' unique contribution will be made by sitting under a tree and thinking all by themselves. The result of that thinking may show up in a later session, at the Evening News, or six weeks later in a company meeting. Open Space requires real freedom, and real responsibility.

Sometimes it happens that overly zealous participants feel that their ideas are so important or powerful that everybody in a particular group (or even in the whole conference) should pay attention and listen. This one has to be nipped in the bud — carefully. The way out is not to directly challenge the person, but rather to remind the assembled group of the Law of Two

Feet. If everybody truly wants to listen, they should do that. But if that is not their desire, they have two feet which they should use. There is no need to argue and shout, just thank the group and leave. Egomaniacs quickly get the picture when everybody leaves.

ADDITIONAL RESOURCES

Leadership Is (170 pages $20 plus S&H) provides an exploration of the nature of leadership in a transforming world, with suggestions on how to do it better. It will be helpful for leaders of an Open Space Event, and to all others who find themselves in a leadership position.

Spirit: Transformation and Development (220 pages $20 plus S&H) offers a broad discussion of the process of transformation in organizations, and what it means to live in Open Space. This is combined with a tested approach for facilitating the process of transformation, along with three major case studies describing what happens when the approach is utilized.

Learning in Open Space (video 25 minutes $65 plus S&H) This video introduces the concept, background, and use of Open Space Technology. It was taped in the United States and India, and follows the two conferences on Learning Organizations.

Open Space Technology: A User's Guide (145 pages $20 plus S&H), a new book presenting everything you need to know about conducting an Open Space event, including logistics, room setup, and a few things to be careful about.

The Open Space Network, an on-line computer conference, is maintained for users of Open Space Technology. For information on this conference or for individual consultation with Harrison Owen, please feel free to call him at the number below.

Supplementary materials, all by Harrison Owen, are available through ABBOTT PUBLISHING. To order, call/Fax 301-469-9269. Or write to:

ABBOTT PUBLISHING
7808 River Falls Drive
Potomac, Maryland 20854
USA